SCIENCE IN A TOPIC
CLOTHES and COSTUME

1400

1564

1700

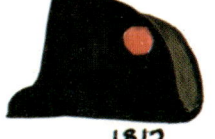

1812

1930

About this book

This book is different from most others because:—

1. It is not complete, but only part of a study — the science part. There will be a need to use many other books to find out about other aspects of the topic — History, Geography . . .

2. It will not tell you information but will only ask you questions and suggest ways that you might find the answers for yourself.
    Many of the suggestions were some children's ways of trying to find an answer — you may have better ideas.

3. It is hoped that arising from these questions other questions will occur to you — do pursue these. (Your own questions and the ways you find to answer them are really the most important.)

4. You do not need to work through the book in the order set out; the sections of work can be done in the order that you wish.

5. There is no need to complete all of one section. If the work becomes harder as you progress through a section, see how far you can go.

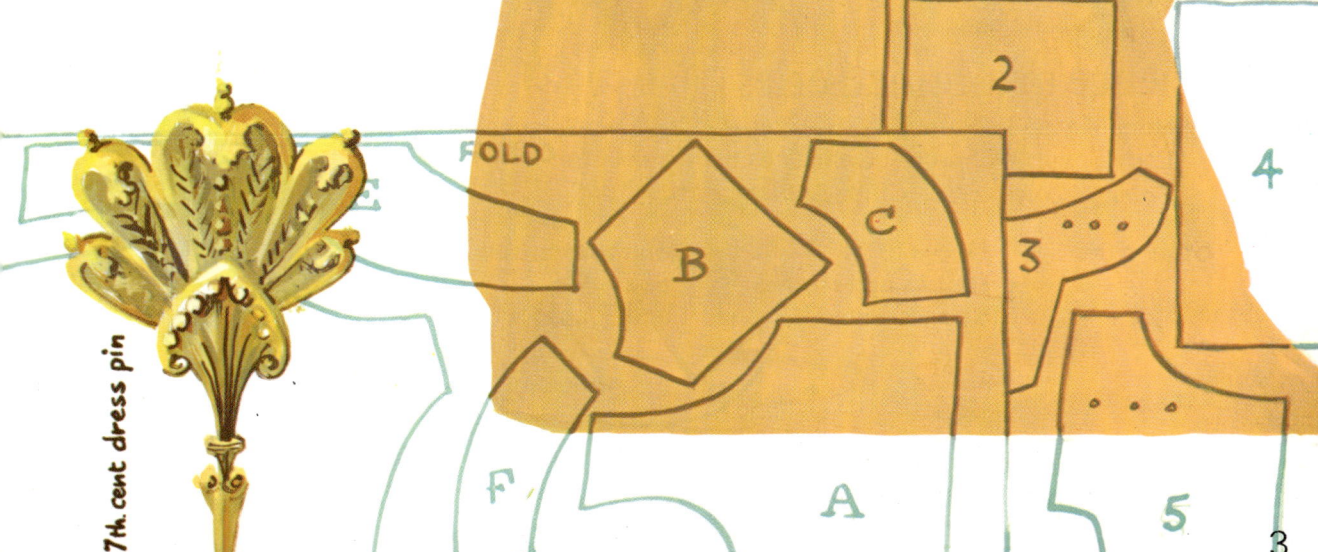

17th. cent dress pin

CONTENTS

Science in a Topic Series

by Doug Kincaid, County Staff Advisory Teacher, Science and
Lady Spencer-Churchill College Buckinghamshire.
Peter S. Coles, B.Sc., Chief Adviser, Berkshire.

Other titles   Ships   Communications
Houses and Homes   Food
Bridges   Land Transport

Note: Panels of this colour with rounded corners represent rather more advanced
work.

LOOKING AT CLOTHES

What differences can you find in clothes worn by people of other countries and other times?
What reasons did these people have for their choice of clothing?

Assyrian   Ballet Dancer

Victorian   1970 Bikini

Above centre:
14th century
English Jester
and an Indian

left: Suit of 15th
century
armour

SECTION ONE

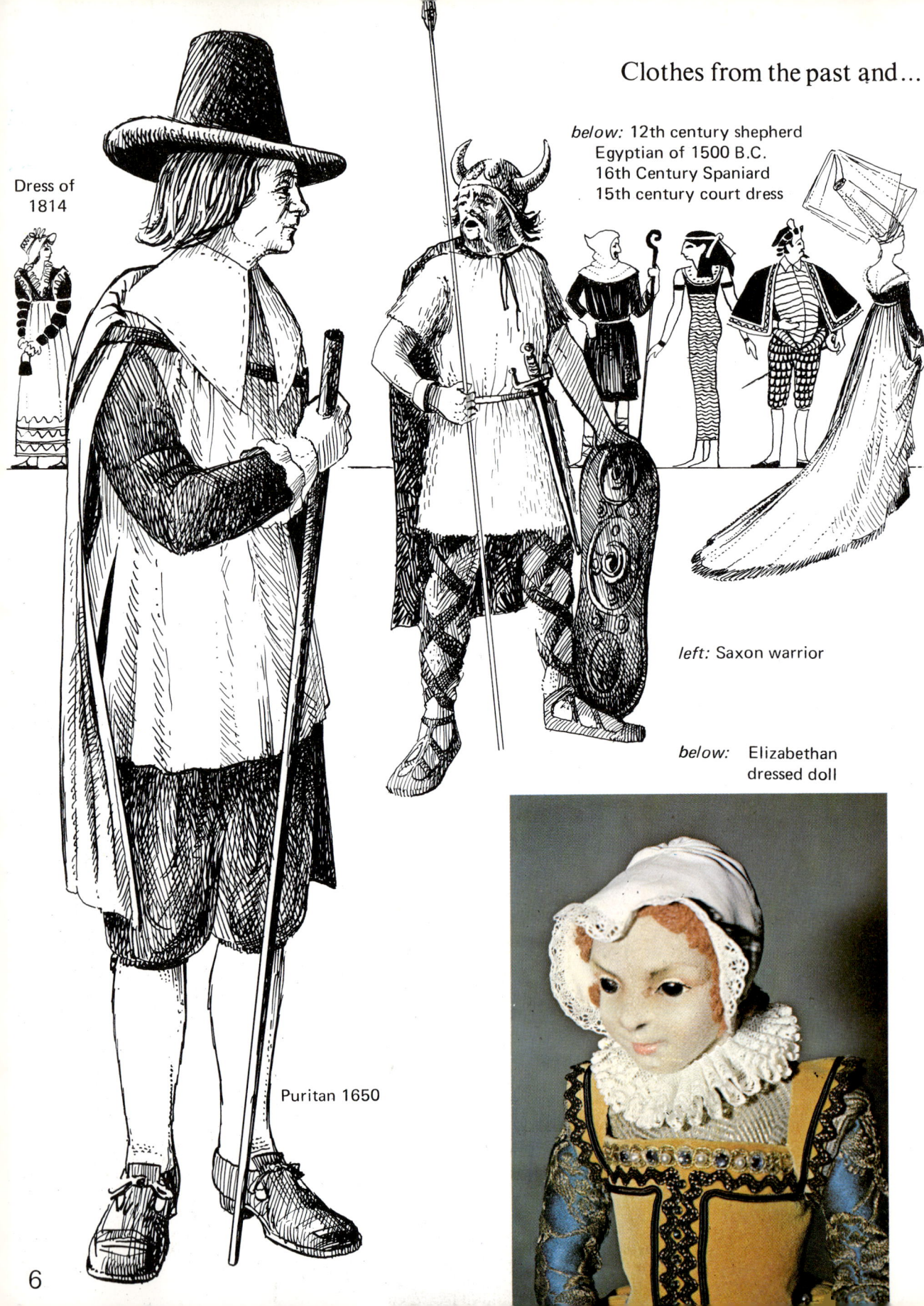

Clothes from the past and...

below: 12th century shepherd
Egyptian of 1500 B.C.
16th Century Spaniard
15th century court dress

Dress of 1814

left: Saxon warrior

below: Elizabethan dressed doll

Puritan 1650

...around the World

left to right: Chinese cheongsam
Mexican poncho
African dancer's beads
Eskimo furs

Why do you think the people on
these pages are dressed like this?

above: Lapps dressed
in traditional
costume

left to right:
Arabian veil & jewellery
Javanese sarong
Indian sari
Spanish clothing

Look at the ways
Special Clothes
are used

Why are these clothes being worn?

above - left to right:
mountaineer
baseball players
speedway rider
cricketer.

left: Fireman in asbestos
suit; industrial
worker with protective
hood; a miner.

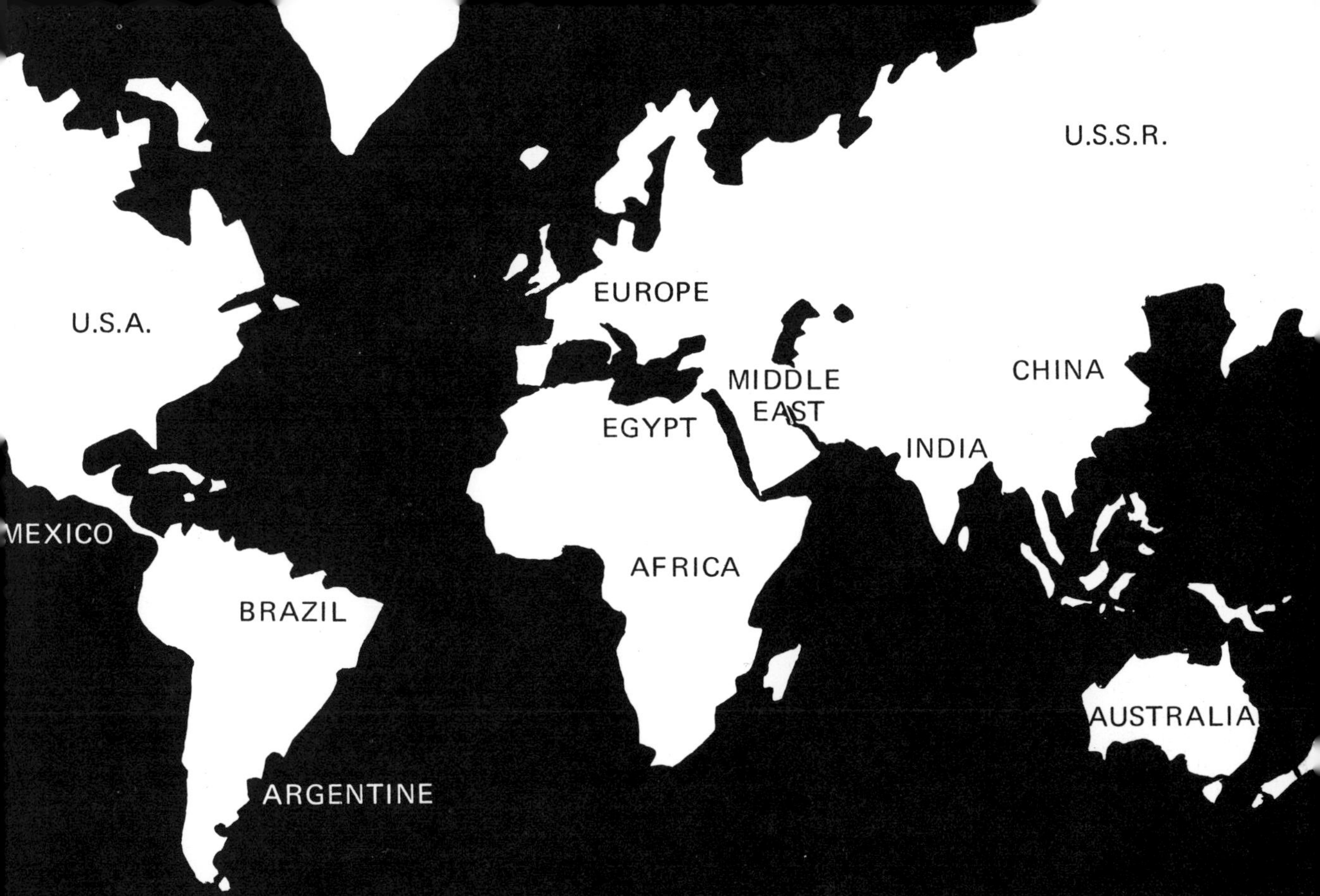

U.S.S.R.

EUROPE

U.S.A.

CHINA

MIDDLE
EAST

EGYPT

INDIA

MEXICO

AFRICA

BRAZIL

AUSTRALIA

ARGENTINE

THE RAW MATERIALS

The raw materials for our clothes come to us from all over the
world. Can you match the sources of raw material with the
places named on the map?

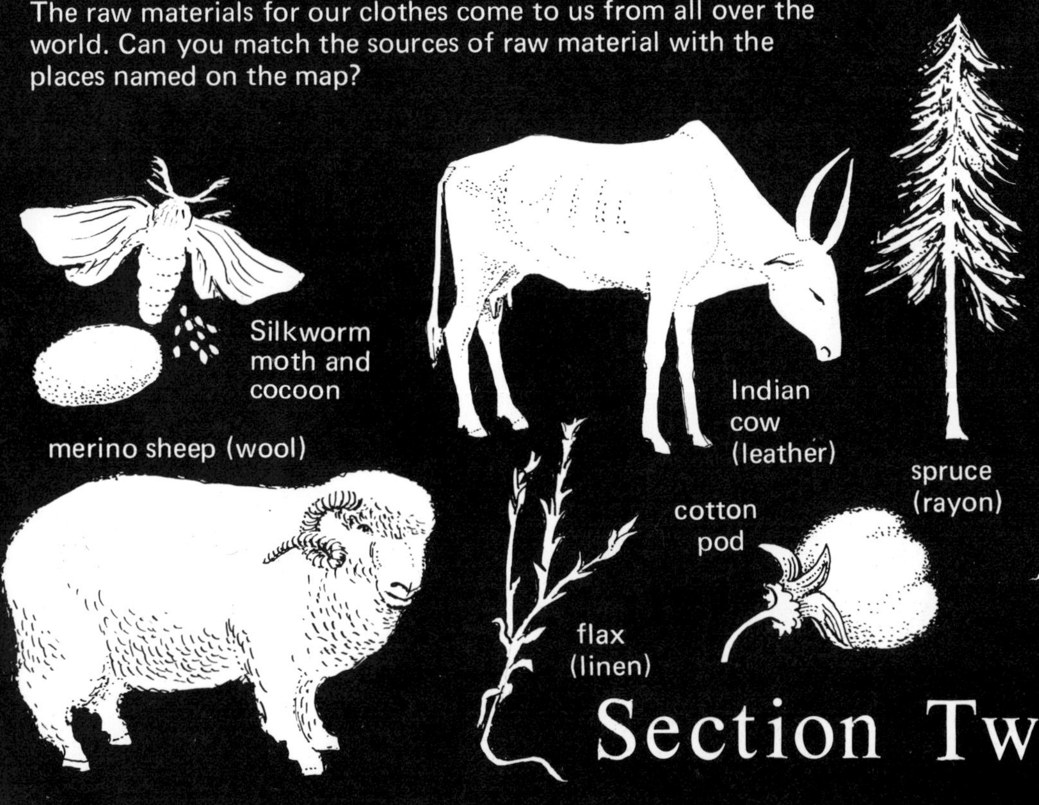

Silkworm
moth and
cocoon

Indian
cow
(leather)

merino sheep (wool)

spruce
(rayon)

cotton
pod

flax
(linen)

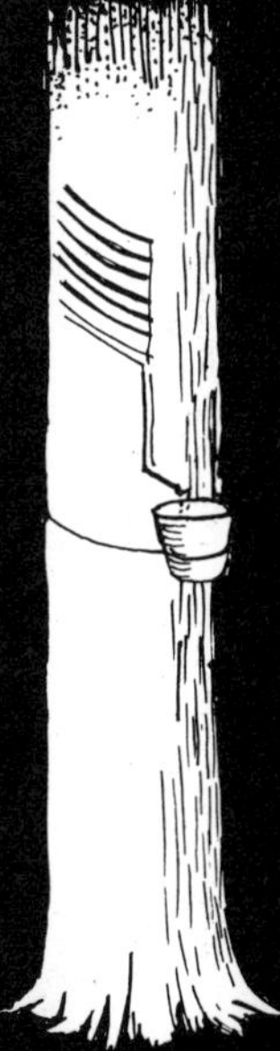

rubber
tree

Section Two

WOOL
From Sheep to Overcoat

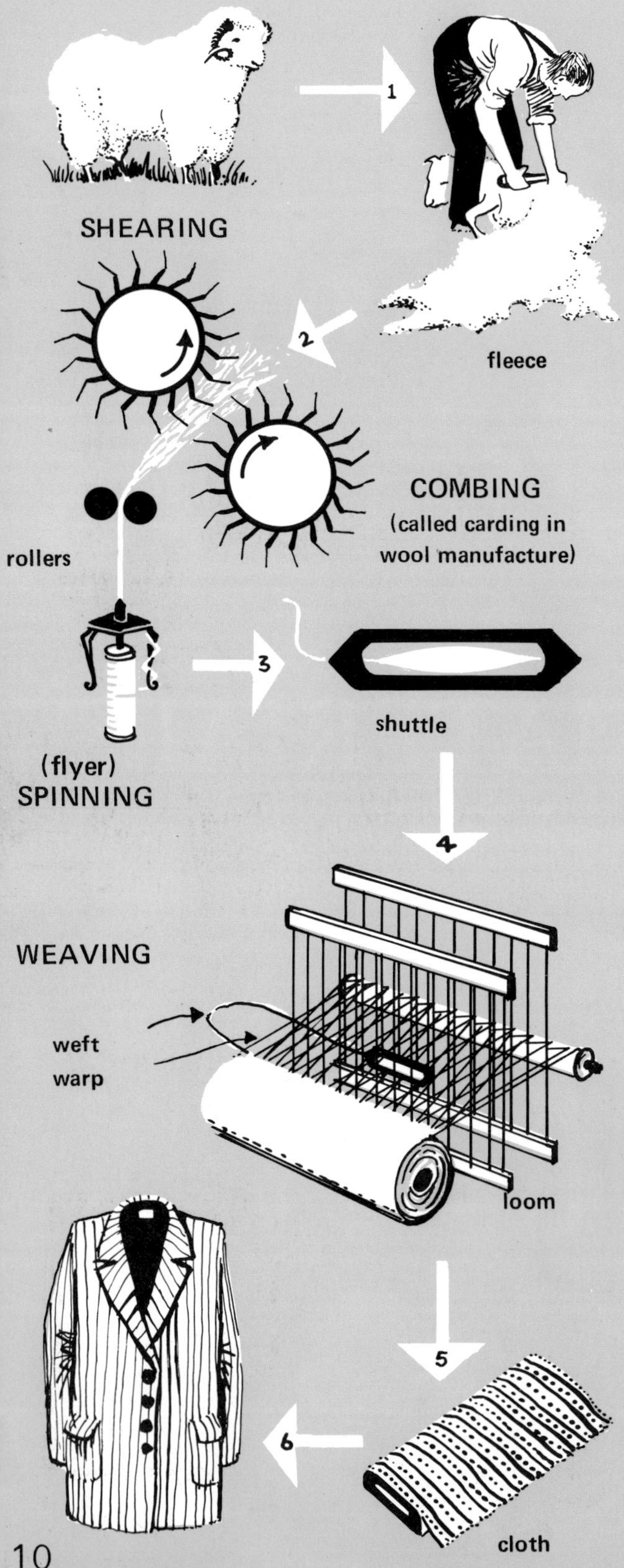

SHEARING

1

fleece

2

COMBING
(called carding in
wool manufacture)

rollers

3

shuttle

(flyer)
SPINNING

4

WEAVING

weft
warp

loom

5

6

cloth

These things are claimed for wool:-
1 Wool is soft and warm to touch
2 Wool will stretch and give
3 Wool keeps us warm

Can these claims be tested?
1 If wool is warm and soft, can it be
  identified just by feeling?
2 How far will wool stretch without
  breaking?

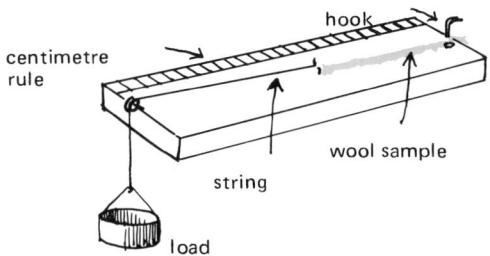

centimetre
rule

hook

wool sample

string

load

After stretching does it go back to
its original length, or has some
permanent stretch resulted?
Does wetting affect this stretch?
Try different plies of wool.
3 If our skin is damp and the damp-
  ness evaporates, we feel cooler.
  Evaporation always causes cooling.
  You could experience this.

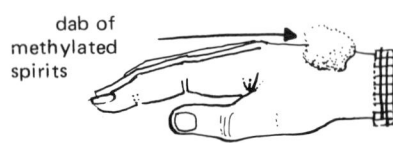

dab of
methylated
spirits

This cooling could result in a chill.
Wool can hold moisture and prevent
such rapid evaporation. Investigate
this with a woollen garment. Dry the
garment thoroughly, find its mass.
Leave the garment in a damp cloak-
room for a morning, check the mass
again and compare.

What clothing is made from cotton?
Which type of clothing is made from these cotton materials?

calico, gingham, towelling, poplin, winceyette, fustian, muslin, corduroy, cretonne, piqué, organdie, flannelette, moleskin and velveteen?

See if you can collect samples of the materials and pictures to make a display or scrapbook.

Your work could include some historical research.

What contribution did these inventors make to the development of the cotton industry?
a John Kay
b Richard Arkwright
c James Hargreaves
d Samuel Crompton

Can you find out why the cotton industry became centred in Lancashire?

COTTON
From Plant to Shirt

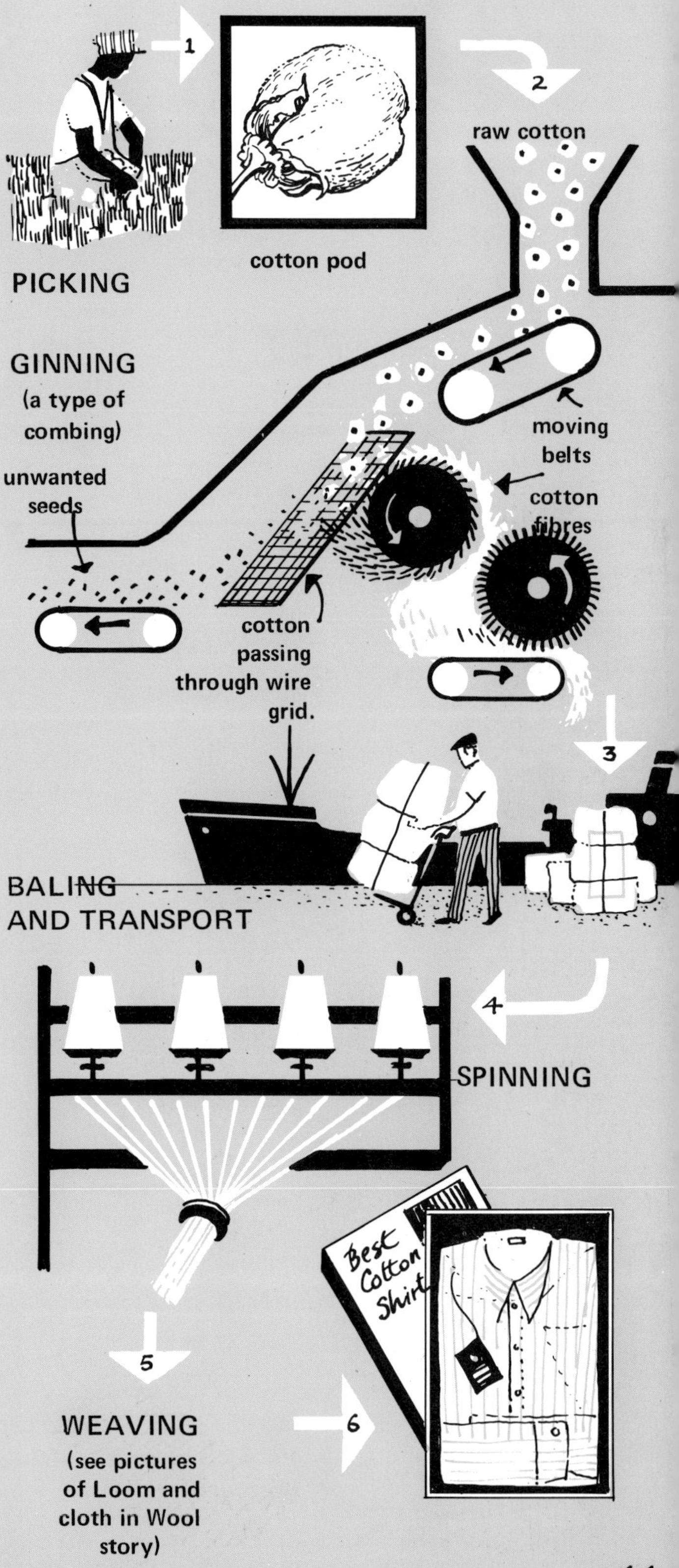

1

cotton pod

PICKING

2

raw cotton

GINNING
(a type of combing)

moving belts

cotton fibres

unwanted seeds

cotton passing through wire grid.

3

BALING AND TRANSPORT

4

SPINNING

Best Cotton Shirt

5

WEAVING
(see pictures of Loom and cloth in Wool story)

6

SILK - From Silkworm to Scarf

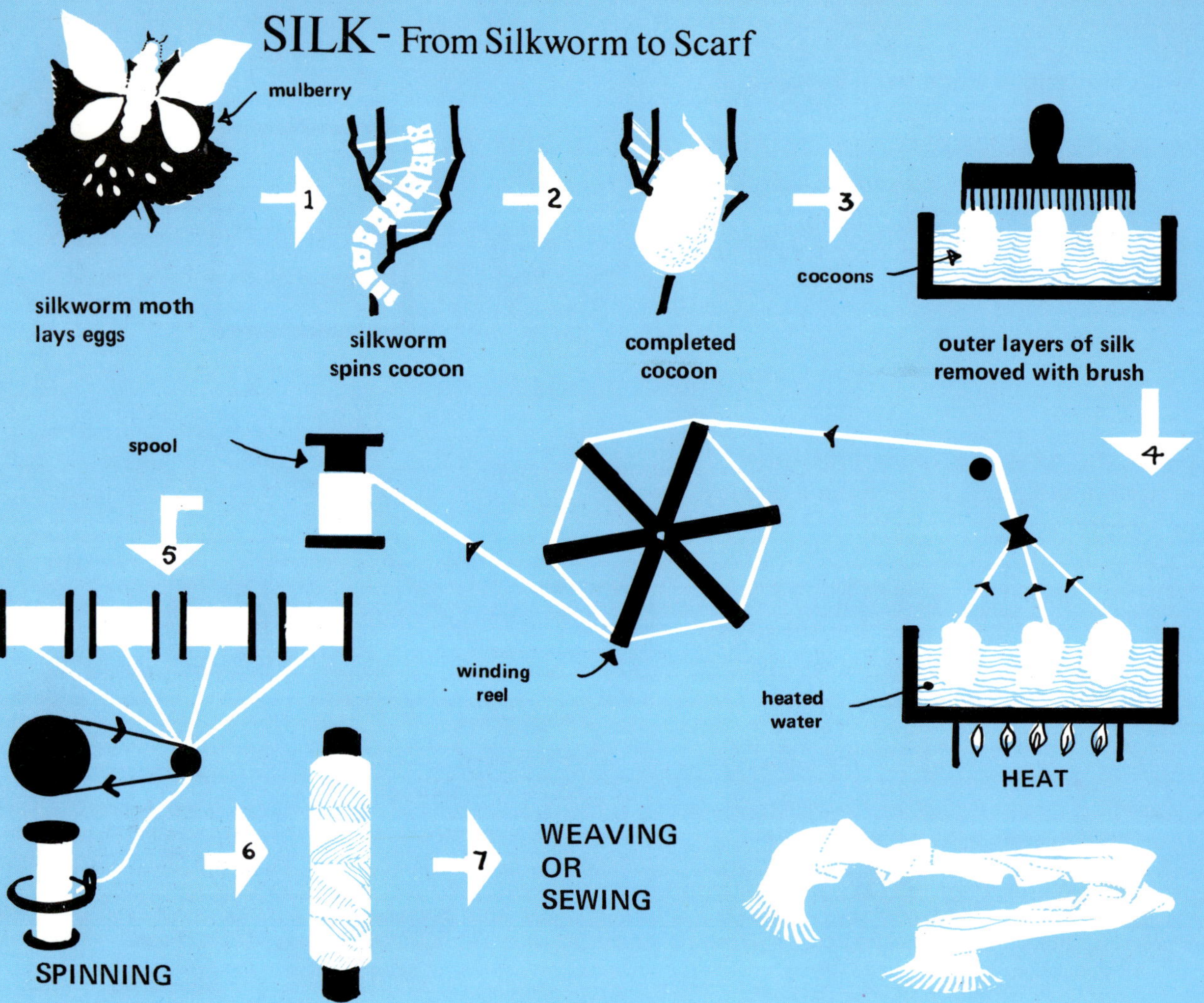

mulberry

silkworm moth lays eggs

1 → silkworm spins cocoon

2 → completed cocoon

3 → cocoons — outer layers of silk removed with brush

4 ↓

heated water — HEAT

spool

5 ↓

winding reel

SPINNING

6 → WEAVING OR SEWING

7 →

You can rear silkworms for yourself and watch the fascinating life cycle of the silk moth.

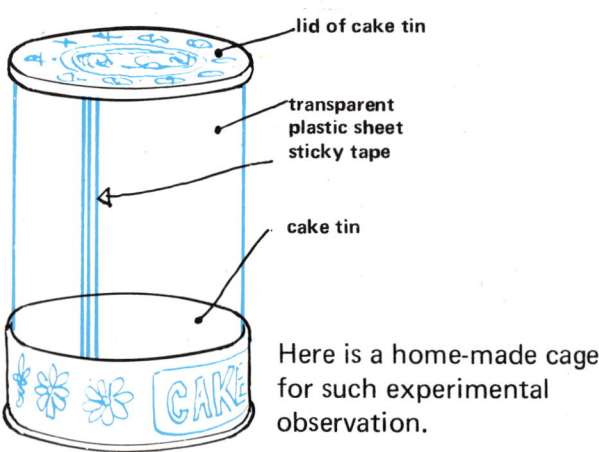

lid of cake tin

transparent plastic sheet
sticky tape

cake tin

CAKE

Here is a home-made cage for such experimental observation.

This man's life was saved by a silkworm. Notice his tie pin. He is a member of the 'Caterpillar Club'. What does this mean?

LINEN-From Flax to Jacket

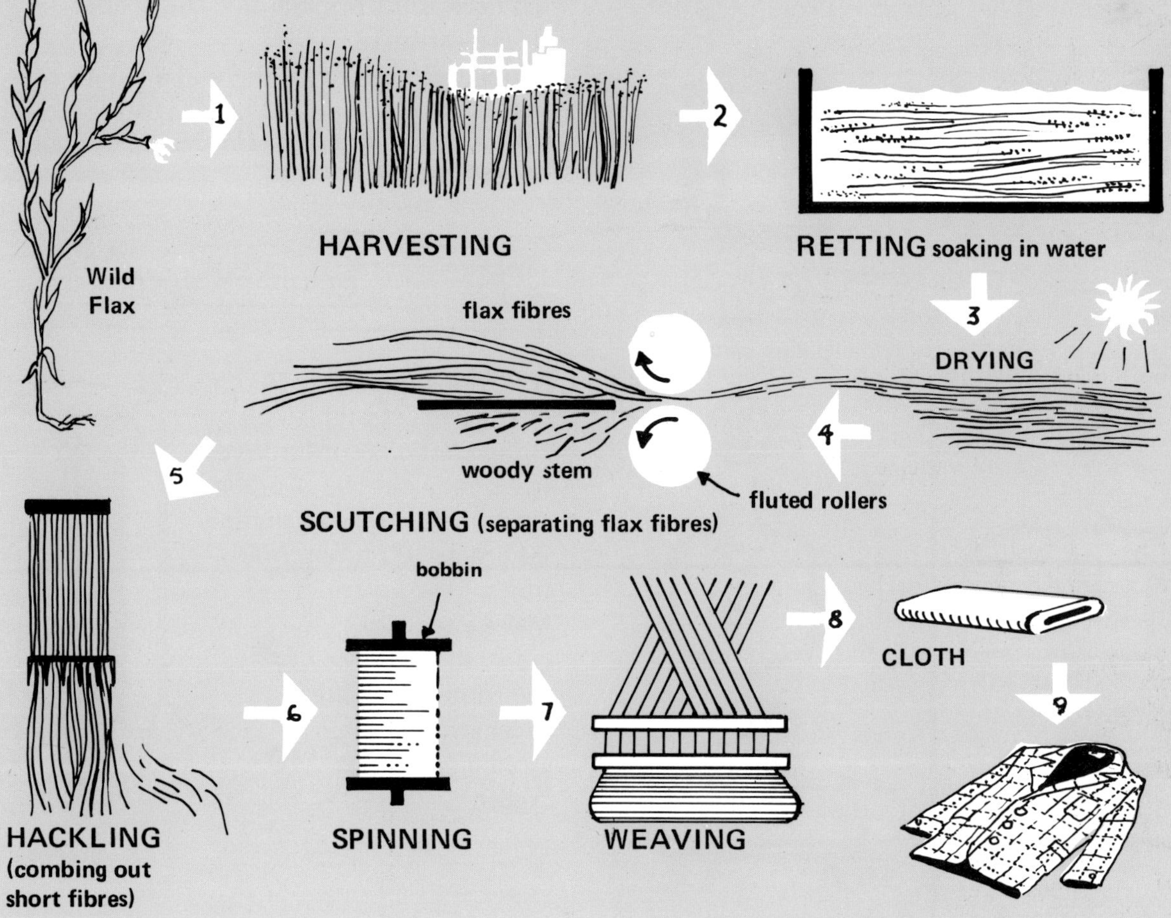

Wild Flax

1 → **HARVESTING**

2 → **RETTING** soaking in water

3 ↓ **DRYING**

flax fibres

woody stem

fluted rollers

SCUTCHING (separating flax fibres)

4 ←

5 ↙

bobbin

HACKLING (combing out short fibres)

6 → **SPINNING**

7 → **WEAVING**

8 → **CLOTH**

9 ↓

This fibre is obtained from a plant stem. Use your microscope to examine some plant stems, and find out about their structure.

The flax is soaked in water and rotting helps to separate the useful fibres. This is called retting. Collect some plants which you think might provide fibres and try retting for yourself.

Jute, hemp, sisal, ramie, kapok, pineapple and coir palm are plants which are valuable because of the fibres they provide. Make a collection of such fibres to examine and compare again using your microscope. The collection plus pictures will make an interesting display.

A large tropical plant like a cactus gives us fibres

13

RUBBER From tree to Foam Rubber or Mackintosh

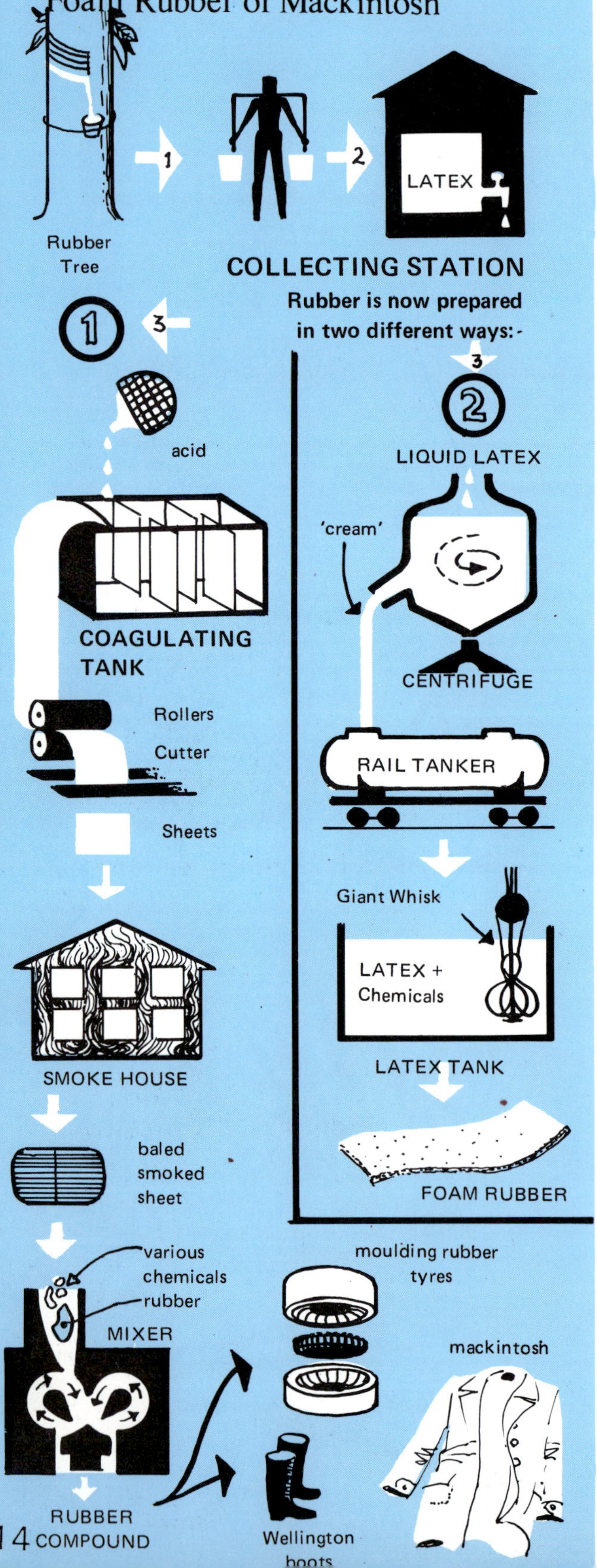

Rubber Tree

1

2

LATEX

COLLECTING STATION

Rubber is now prepared in two different ways:-

① 3

3

②

LIQUID LATEX

acid

'cream'

COAGULATING TANK

CENTRIFUGE

Rollers

Cutter

Sheets

RAIL TANKER

Giant Whisk

LATEX + Chemicals

SMOKE HOUSE

LATEX TANK

baled smoked sheet

FOAM RUBBER

various chemicals
rubber

moulding rubber tyres

MIXER

mackintosh

RUBBER COMPOUND

Wellington boots

Rubber will stretch and then go back to its original size and shape. It is elastic.

How much stretch is there in a length of elastic?

Cut a 20 cm length of elastic. How much does it stretch?

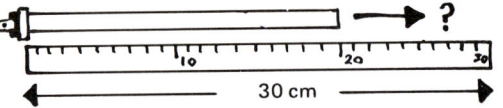

?

30 cm

If 20 cm stretches to 30 cm does 40 cm stretch to 60 cm? If the elastic is continually stretched does the stretching pattern change? Try pulling 20 times, 100 times, 500 times.

Make a table or graph showing how the rubber increases in length as the stretch force gets bigger.

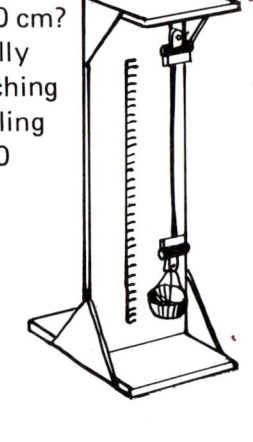

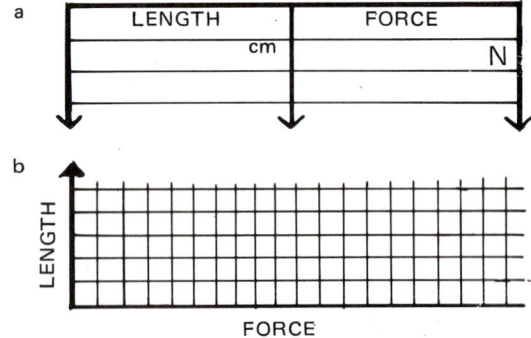

a

LENGTH		FORCE	
	cm		N

b

LENGTH

FORCE

Continue your investigations with different types and widths of elastic and with different forms of rubber by using a test rig as shown.

You might be surprised if you can notice the effect of pouring hot water over stretched rubber you are measuring.

LEATHER

From Hide to Boots
From Skin to Gloves

1 Leather of one kind or another comes from the skins of animals. 'Hide', a thick strong skin, comes from cows, horses and buffaloes. Skins of sheep, goats, pigs and small animals are simply called skins.

SORTING

2 skins

tanning solution

large wooden drum

lift

TANNING

3

drum

Formic acid

dye

skins

DYEING (4 - 6 hours)

roller cutter

skin

5

SHAVING (grained skins)

4

WHEELING
(for suede leathers)

6

'spindryer'

'mangle'

7

skin

'STRIKING OUT'
(removal of moisture by spindrying and mangling)

8

1 VACUUM DRYING
2 IRONING
3 MEASURING
4 FINAL SORTING

9

CUTTING & SEWING

15

RAYON   MAN-MADE FIBRES   NYLON

spruce trees-logs

1 → 2 → MILL

3

4

wood pulp

CAUSTIC SODA BATH
carbon disulphide

5

GRINDING
into pulp

6

churn

7

SPINNING

VISCOSE

spinneret

8

ACID

WEAVING → 9

cloth

WATER   AIR   CHEMICALS FROM COAL

CHEMICAL PROCESSES MAKE RAW NYLON

1

Nylon liquid

Rollers make into sheet

GRINDING

HEAT PROCESS

Melted nylon
extruded
through SPINNERET

2

SPINNING

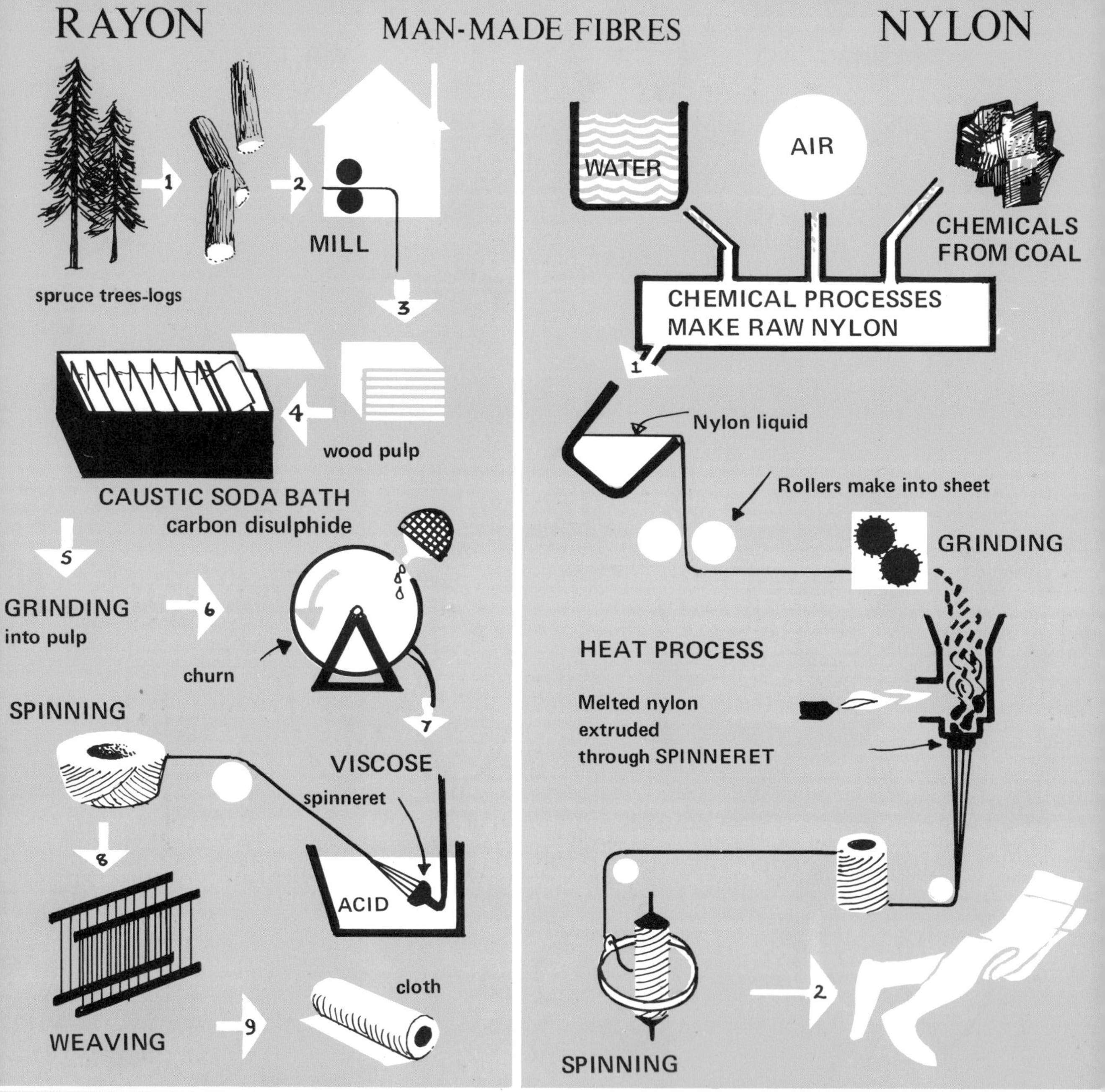

The term 'man-made fibres' is used to describe those created by chemists. Over half the world's clothing now uses these synthetic fibres.

Claims are made that in many ways they are better than the natural fibres. You can carry out some research into this in the next section.

You can experience the process of 'extruding' a plastic fibre with a tube of balsa cement.

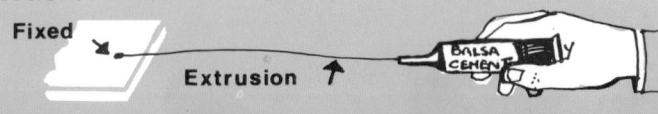

Fixed

Extrusion

BALSA CEMENT

Before leaving this section it would be interesting to search for examples of unusual raw materials being used for fibres. Metals - ? Asbestos - ? Glass - ?

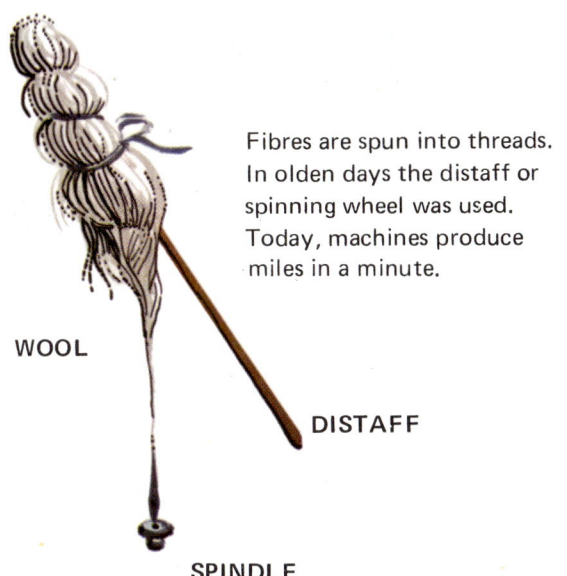

Fibres are spun into threads. In olden days the distaff or spinning wheel was used. Today, machines produce miles in a minute.

WOOL

DISTAFF

SPINDLE

SPINNING WHEEL

Fibres, Threads & Fabrics

Today a machine produces yarn, 50,000 times faster than a hand spinner. Threads are woven or knitted to make fabrics.

Try these processes for yourself and you will find why faster ways were needed.

FIBRES INTO THREAD

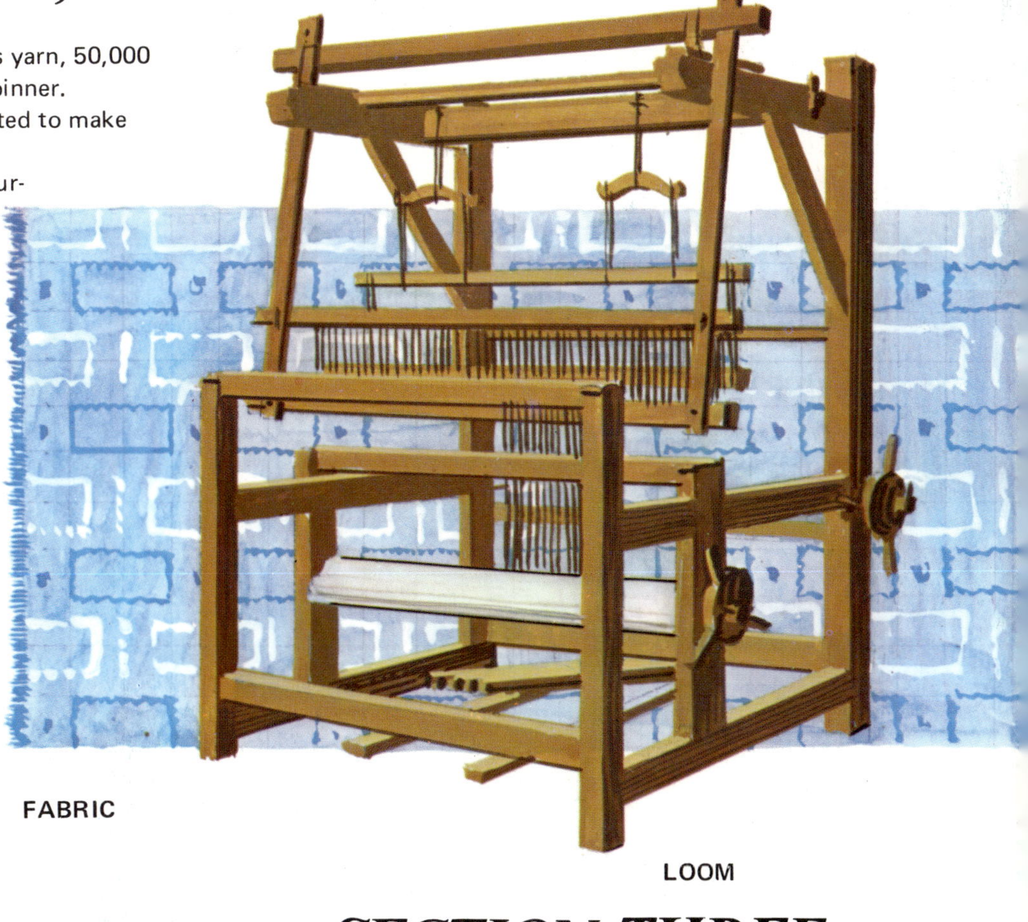

FABRIC

LOOM

SECTION THREE

FIBRES AND THREADS - Identification

Clothing is made from many different fibres. Often a mixture is used, sometimes one type for the warp and another for the weft. As your study develops you will need to know the fibres used.

How can fibres and threads be identified?

One way is to collect a selection of known fibres and threads. These can be bought or collected from fields and farms.

You can test these and build up a record of observations.

1 Look at them through a micro-scope. Draw what you see. Here are one young scientist's drawings. Which threads did he record?

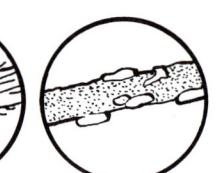  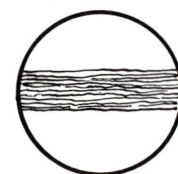

2 Feel them.

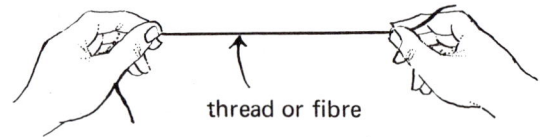

thread or fibre

Hold the fibres between the finger and thumb of each hand, and slowly draw apart. Try one of your own hairs in each direction. Why does the hair slide in the way it does?

3 Try burning a short length, holding the piece with tongs or tweezers.
   a How easily does the sample burn?
   b Does it continue to burn when removed from the flame?
   c If it does not burn, does it melt?

   d Describe the burning or melting, particularly the smell and smoke.
   e Describe what is left.

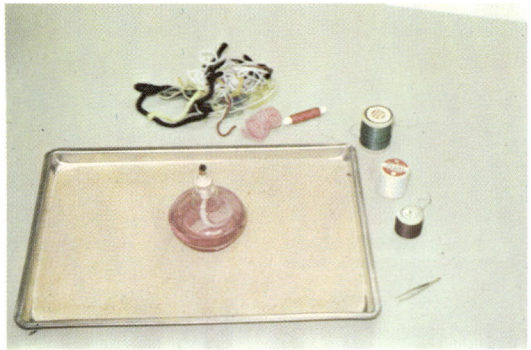

Apparatus: Tray, burner and tweezers

Burning test

Make a booklet about your discoveries

A page for each with a sample, drawings and descriptions will be a valuable reference.

Now try to identify some unknown fibres and threads

SAFETY FIRST — USE AN ASBESTOS MAT OR SAND TRAY

HOW STRONG ARE THREADS?

First measure the thickness of some threads, as this is something you will wish to know when comparing strengths.

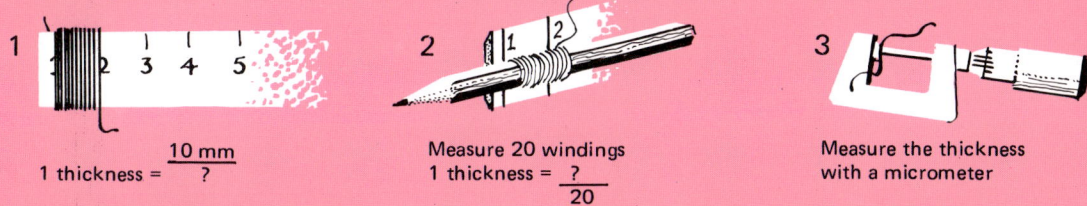

1 thickness = $\dfrac{10\ mm}{?}$

Measure 20 windings
1 thickness = $\dfrac{?}{20}$

Measure the thickness with a micrometer

How can strength be measured?

Here is a thread - breaker made in a school

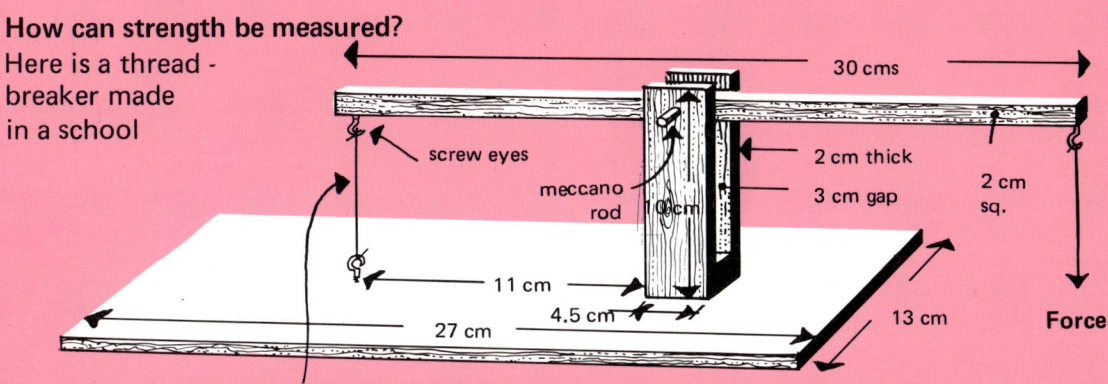

screw eyes

meccano rod

30 cms

2 cm thick

3 cm gap

2 cm sq.

11 cm

4.5 cm

27 cm

13 cm

Force

To keep the test fair think about this length

a What kind of thread is strongest?
b Are two threads twice as strong as one, or perhaps more than twice as strong?
c Does twisting or looping the sample change the result?
d Are wet threads weaker or stronger?

e How does strength relate to thickness?
f Do some stretch more than others before breaking?
g Fishing lines have breaking strains stated; you could check these.

The force needed to break a thread should be measured in NEWTONS. If you used a load to break your threads, now measure its pull using a force meter.

Thread breaker in use with load

Hanging bucket from force meter

STRETCHING

Clothing is stretched in some places more than others. Trousers become baggy. Dresses lose their shape.

A thread that stretches but returns to its original length would be useful for clothing.

How do different threads stretch?

Here is a way of researching into stretching that one school used.

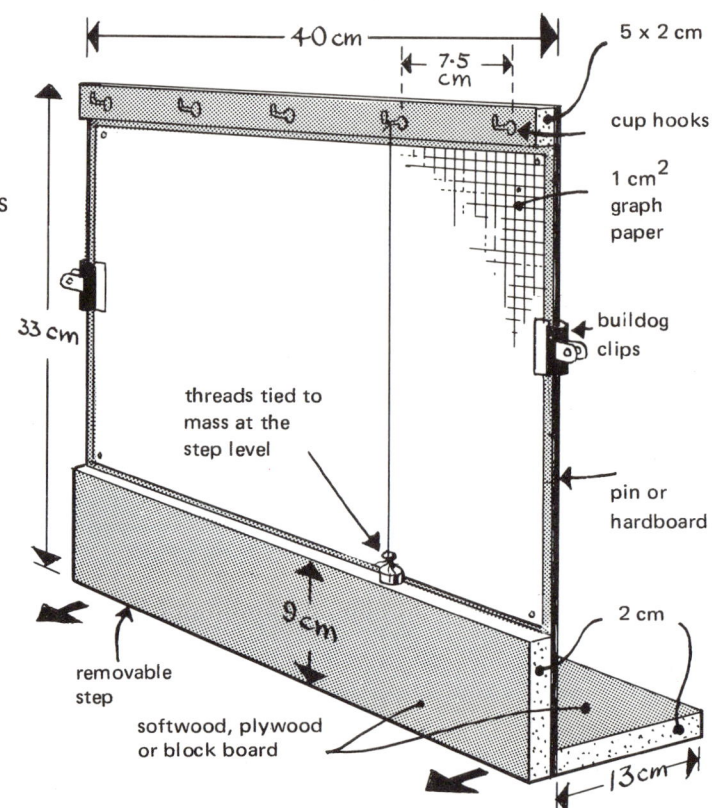

40 cm

7·5 cm

5 x 2 cm

cup hooks

1 cm^2 graph paper

33 cm

threads tied to mass at the step level

buildog clips

pin or hardboard

9 cm

2 cm

removable step

softwood, plywood or block board

13cm

right: A stretching test rig

Preparing the experiment - tying a thread to mass before removing step.

Recording the results of the experiment after an hour.

a How does stretch vary with time?
b What difference does wetting the thread make?

c How much are different threads permanently changed?

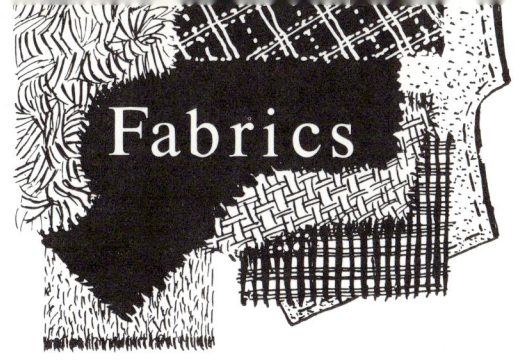

Fabrics

Looking at Cloths

To investigate the fabrics used for making clothing, start by making a collection.
Sources for such a collection could be - mother's scrap bag, needlework catalogues, a dressmaker and old pattern books from shops.

You will be using these for experiments but do keep a sample of each for display.

How much can be found out about each sample?

a Is it woven, knitted, felted or bonded?

b How is it given its colour? Is it dyed or printed?

c How is it patterned? Is it printed, woven or textured?

Try sorting by feeling them.

Examine the sample with your microscope. Try pieces of fabric between slide cover glasses and project

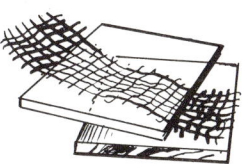

Place fabric between slide cover glasses.

Making a cloth count

The weaver talks about the count of a cloth. This is the number of warp threads and the number of weft threads per unit length.

Find the cloth count of one of your samples. Is there a connection between cloth count and use?

You might like to experience making fabrics from threads by knitting or weaving.

Here are some weaving patterns.

Plain

Twill

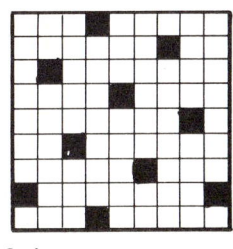

Satin

Strength and Stretch

Examine the threads used in making a fabric.
Are the warp and weft threads different?

Perhaps they are different kind of fibres.
Perhaps different thicknesses of threads are used.

How can different fabrics be tested for strength and stretch?

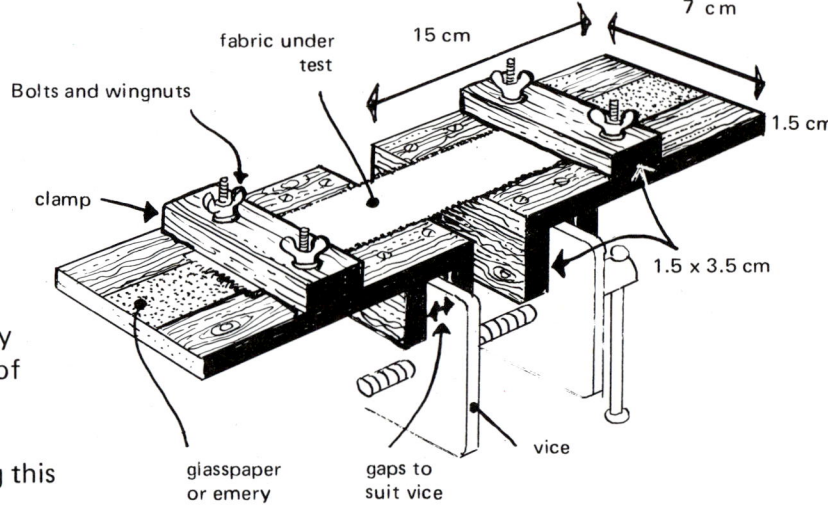

The forces needed can be very large, so use a narrow width of fabric (2 cm).

Here is one way of measuring this strength and stretch.

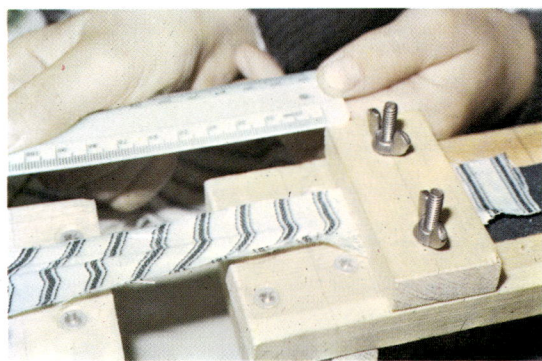

Recording stretch

Try cutting your test piece from at least three directions.

Compare different weaves, knitted and woven.

How does cloth count relate to stretch?

How do mixed yarn fabrics compare with single yarn fabrics?

Torn fabric

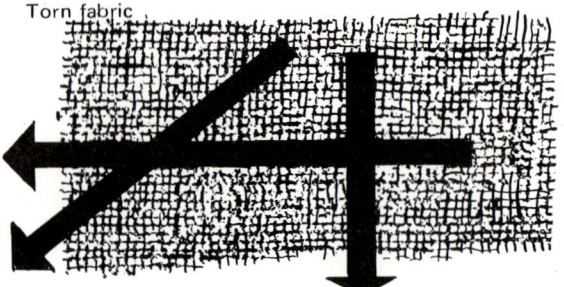

FIBRES AND THREADS Threads are used to join fabrics and make clothes. Pretend you are stranded on a desert island. Can you join some materials together to make a simple foot covering or another small item of clothing? Some modern materials like plastics are joined with adhesives or by 'bonding'.

What other ways of joining fabric can you find.

Keeping Warm and Dry

Why wear clothes? One reason is to keep warm How are these people keeping warm?

Another reason is to keep dry.

What are these items of wet weather clothing called? Can you find out why they are so named?

SECTION FOUR

23

STAYING WARM

How good at keeping us warm are the various types of clothing?

Which is the best way of keeping warm?

a one thick layer?

b several thin layers?

c loosely woven material?

d heavy blanket type material?

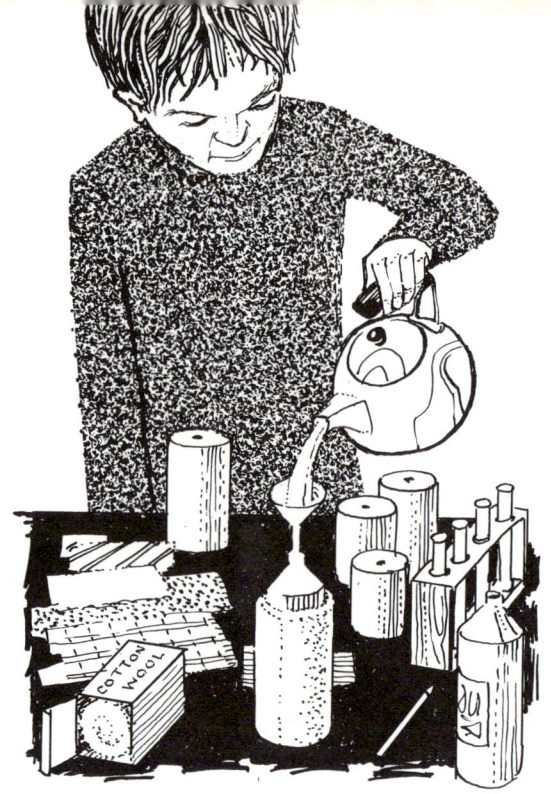

Here is an experiment that will help to find the answers.

You need a squeezy bottle, tin or test tube filled with hot water.

This represents your warm body.

How can the heat be stopped from escaping too quickly?

Wrap 'the body' in the various materials under test.

(leave one unwrapped - why?).

Experiment with:-

a One layer of blanket

b Two, three, four layers of blanket.

c Several layers of thin material.

d Aluminium foil.

e A string vest plus a thin layer.

f Cotton wool

g Layers of paper

h Different fabrics

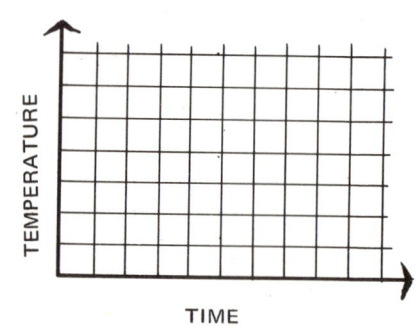

Record your thermometer readings and draw a cooling curve graph.

LOSING WARMTH

How does wind affect the loss of heat?

Try this experiment:-
    Place two lengths of bandage in hot water.
    Wrap these around thermometers.
    Note the temperatures.
    Hold one in front of a fan.
    Take further temperature readings of each.
You could show this more dramatically by noticing the cooling due to the evaporation of other liquids.

You may have read of people trapped on mountains and dying from exposure. This is due to loss of body heat, not only from the cold but because of wind.

You could test how windproof a material is by using a candle and a fan.

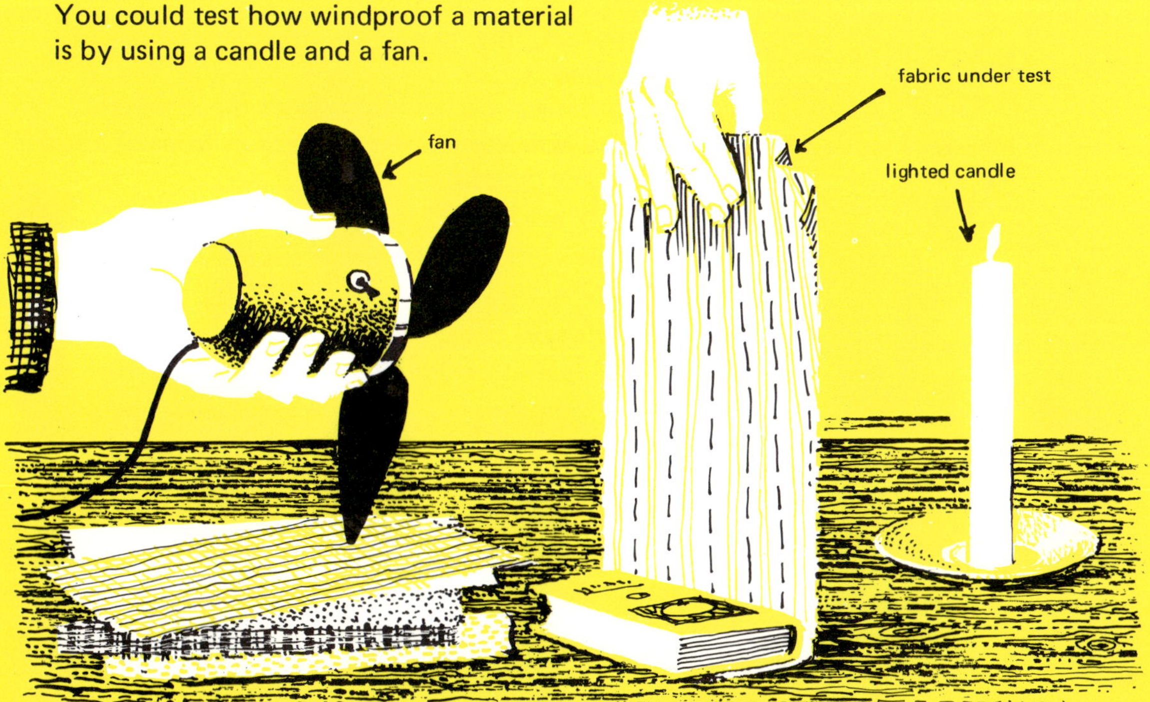

fan

fabric under test

lighted candle

Which fabric keeps all wind from the candle flame?
With which does the flame only move slightly?
Which offers least protection?

KEEPING COOL

Is white clothing cooler than dark coloured?

Experiment with some painted tins.

Fill them with hot water
Let these represent you,
full of warmth.

Does one lose heat
faster?

Fill with cold water.
Does one gain heat
faster when placed
before a fire?

IDENTICAL TINS

painted	painted
matt	matt
black	white

Stand on paper or card

How do the coloured clothes we wear affect how warm we feel?

You will need to have material of the same kind and thickness, the only difference being colour.
Your school will have some felt squares used for needlework that are exactly the same except for the varied colour.
Try placing a thermometer under the different colours with the squares arranged in front of a fire or outside on a sunny day.
Can any variation in temperature be found?
There is a new lining material for clothes that reflects back your body heat.
Try this idea for yourself.
Hold a piece of aluminium foil in front of your face.

These people seem to have chosen white or light coloured clothes to wear. Why?

Time how long it is before a drop seeps through

KEEPING DRY

Which materials keep out water?

Investigate how different fabrics behave when water is dripped on to them.

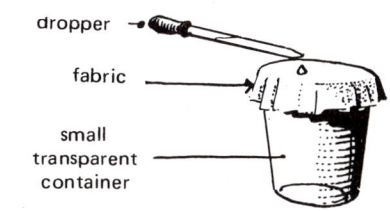

Sort the fabrics into sets.
  a Which let water through quickly?
  b Which does the water soak through, after a time?
  c Which does the water run off?
Which material would be good for a raincoat?

Plastic might seem the best, but does it have disadvantages?

How does this experiment compare with you in a plastic raincoat?

Try an experiment like this

How much water do different materials *absorb?*

Take squares (15 cm x 15 cm) of materials. Place the squares in water, remove and shake off the drops.
How much water is absorbed?

As your piece of cloth is quite small the water absorbed will not be very great. You may therefore have difficulty in measuring such a small mass. You will need to detect ten grams in one gram steps. If your school does not have such a sensitive balance, you may like to try your inventive skills to make such a balance.

Here are some suggestions to help you start:-

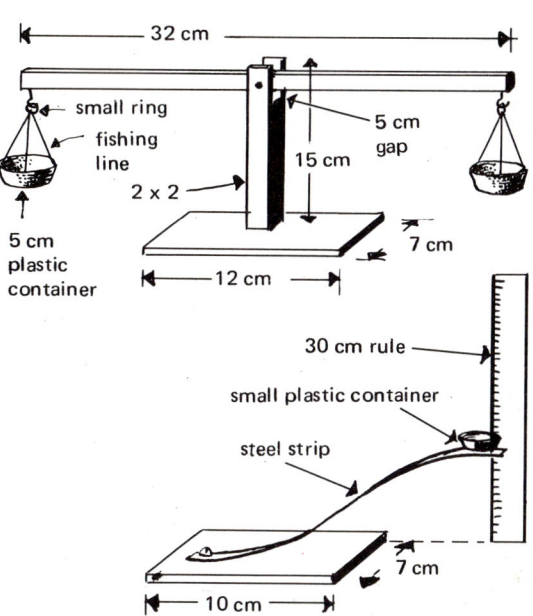

SOAKING WET

How do different fabrics absorb water?

You could find out with an arrangement like this:-

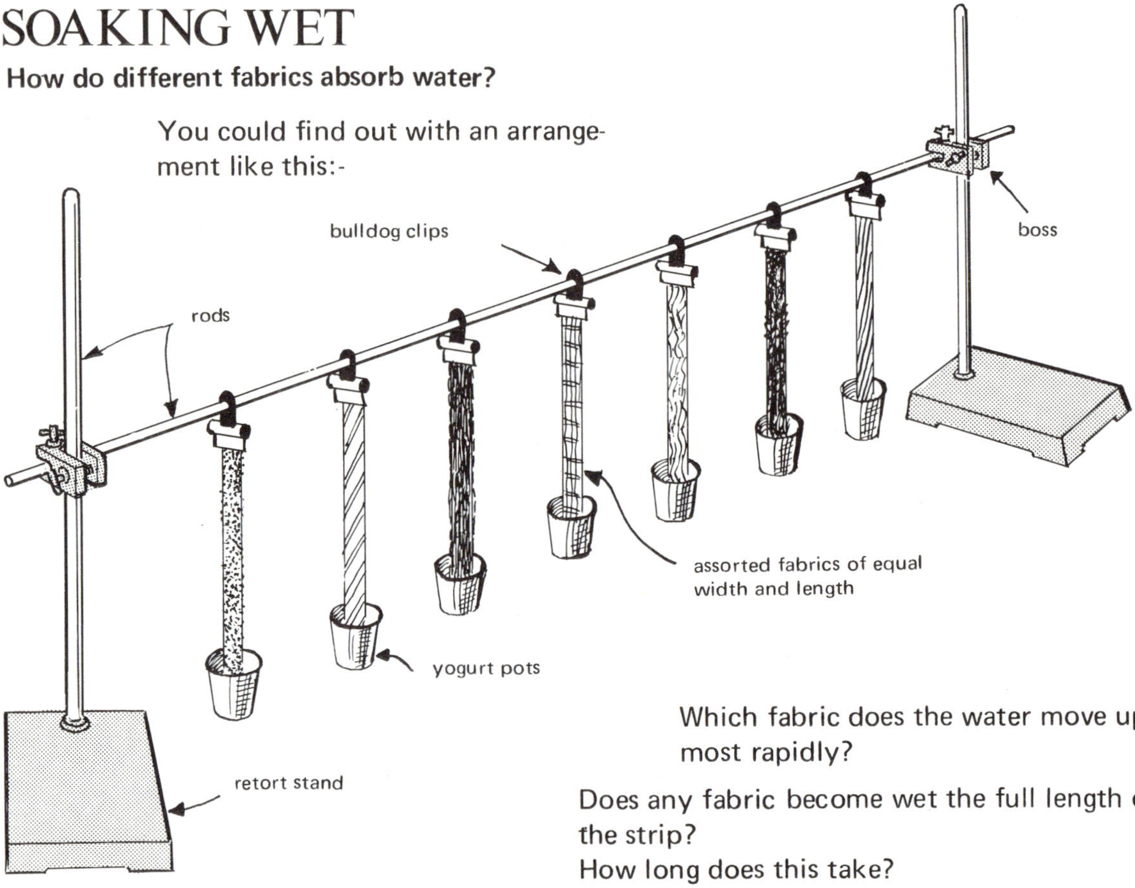

bulldog clips

boss

rods

assorted fabrics of equal width and length

yogurt pots

retort stand

Which fabric does the water move up most rapidly?

Does any fabric become wet the full length of the strip?
How long does this take?
What difference does cloth count make?

Is there a difference if the strip is cut in different directions?

You could extend this research to other materials.
It would be interesting to experiment and find the best paper towel for soaking up water.

Experimenting with paper towels

Measuring mass of water absorbed

Who needs Special Clothing ?

Most special clothing has been designed to protect people.
What are these people doing?
What do they need protection from?

SECTION FIVE

29

PROTECTION
Bites and Bumps

What are these people trying to protect themselves from?
In this country it could be an unpleasant sting. In some lands overseas it could be a bite that caused a fatal disease.

Seeing what it is like under a mosquito net

How are these people trying to protect themselves from bumps and blows?

A soldier of the past. Does a modern soldier use such protection?

An American footballer. Which other sportsmen wear protective clothing?

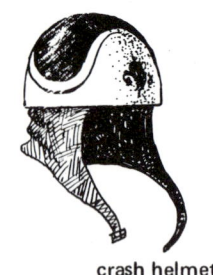

crash helmet

riding cap

Why are helmets this shape? Who else uses a helmet to protect his head?

Dust and Dirt

Overalls, aprons and coats are used as protection from dirt.

Experimenting with various materials

Which materials are best at protecting us from dirt?

Collect some materials that could be used for dirt protection.

How well do they protect?

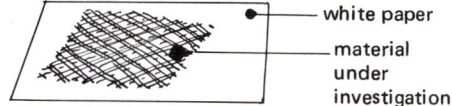
white paper
material under investigation

Rub on some dirt, grease, black paint powder. Does the material protect the white paper?

Which is the best type of cloth to use for protective garments?

Here are some facts and figures about injuries in one year caused by people not wearing the correct protective clothing.

79 000 Hand Injuries
44 500 Foot Injuries
13 500 Head Injuries
11 000 Eye Injuries

What protection is needed?   Here is some research for you:

Person	The protection needed	Item of clothing	Material made from
Fireman	From fire and heat	Fireproof suit	Asbestos
Diver			

PROTECTIVE CLOTHING An important type of protective clothing helps to stop harmful amounts of radiation. For example, in hospitals, radiographers wear lead aprons when working with X Rays. Other workers need protection from Gamma Rays and radioactive particles. Can you find pictures of workers wearing this type of protection?

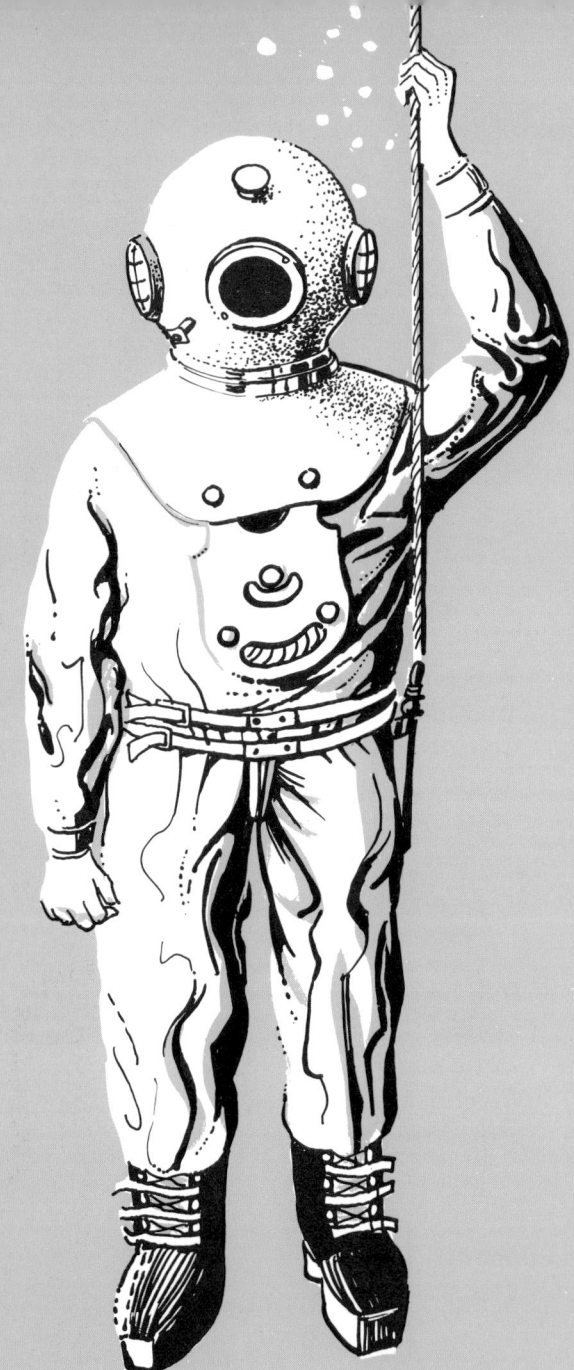

THE DIVER

a Why does this diver need a mass of lead attached to his chest?

b What are the various parts of his suit made from?

c How does he breathe when beneath the water?

d What sort of pressures must he with-stand?

e Why does he feel different when moving:
   i on the land?
   ii in the water?

f How will he communicate with the surface?

g Why are his boots so heavy?

h How is the suit made to allow for movement?

What is the difference in forces acting upon him?

Carry out some investigations that will help provide some answers.

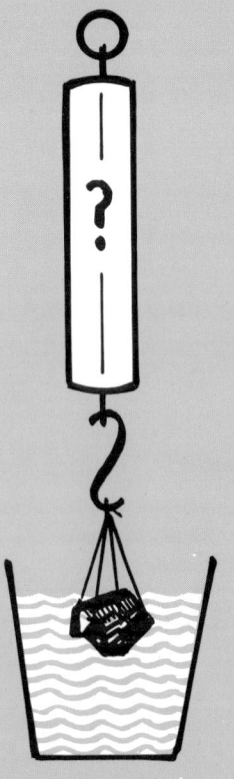

in the water?

on land?

Find out how the water presses
(work outside)

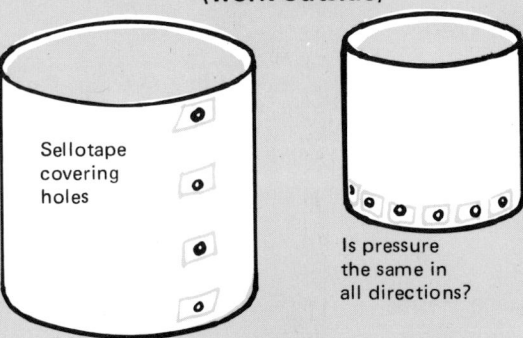

Sellotape covering holes

Is pressure the same in all directions?

You need a large tin with holes drilled and covered with masking tape. Fill the tin with water, remove hole cover and observe the streams and distances.

Can you predict where to place the catch can?

Will there be differences if the tin is only half filled?

What does this tell us about the diver and pressure at different depths?

Is the pressure only downwards?

Use a clear plastic tube with a piece of card held over the bottom by a thread. Push this into the water and find out what force is needed to push the card away.

This could be done by pouring water into the tube.

The Cartesian Diver

A scientific toy which can be made from a medicine dropper and a bottle gives rise to many intriguing questions.

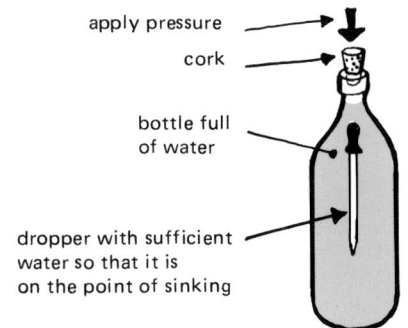

apply pressure

cork

bottle full of water

dropper with sufficient water so that it is on the point of sinking

Why does it sink and float as the pressure is varied?

What does this tell us about the ease or difficulty of 'squashing up' liquids and gases?

This home-made manometer could help you to measure *pressure changes*

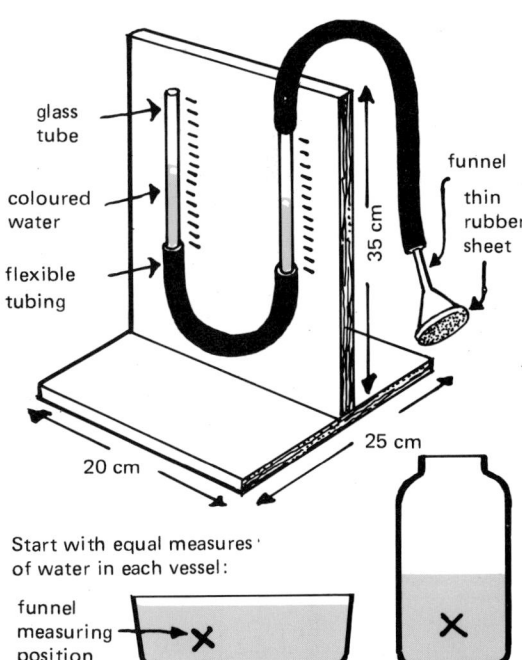

glass tube

coloured water

flexible tubing

funnel

thin rubber sheet

35 cm

25 cm

20 cm

Start with equal measures of water in each vessel:

funnel measuring position

X

X

Is pressure the same at the same depth?

Is pressure the same at the bottom of each container?

Fill the containers and measure pressures.

UNIFORMS

Another special type of clothing is the uniform. Who wears a uniform?

What can you find out about the origins of the yeoman warder's and Vatican guard's uniforms?

Why does the sailor wear bell-bottom trousers?

Think of reasons for and against school uniforms.

Why does the nurse wear white starched cotton?

Who has a uniform that tells us he rides a horse?

A security guard ·

left to right: Yeoman Warder; Vatican Guard; Canadian mounted policeman, Sailor, Policeman, Schoolboy

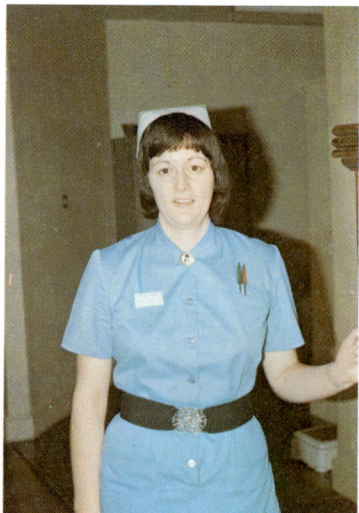

A nurse

Air hostesses

THE SOLDIER

Look at a soldier's uniform:

When does the soldier wear each of these kinds of clothing?

Here is a toy soldier

'Action Man'

Why is his battledress daubed with green and brown?
What is 'Action Man's' uniform for the Arctic and
for the desert?

**This colouring and pattern designed to merge with
a background is called camouflage.**

What else is camouflaged?

Try using 'Action Man' or a model tank or plane
and painting a background for it. Who can do the
best 'disappearing trick'?
How could you measure a successful camouflage?

HIDE AND BE SEEN

Camouflage is very important in the animal world. It can mean escape from enemies and therefore survival.

Above: Praying Mantis
Right: Bullhead Fish

How do animals 'clothe' themselves to hide?
a Which uses colour?   c Which uses pattern?
b Which uses shape?

How do these animals fit into their backgrounds?

 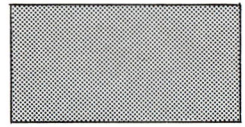

Investigate camouflage for yourself

Paint and cut out, or model some creature — a fish, a bird or a mammal. Paint a background for it. How well can you make it 'disappear'?

Can the success of your camouflage be measured?
Perhaps the distance at which it can be detected will provide a scale. Perhaps more than one creature could be used and the measure be the number that can be seen in a few seconds.

If you try moving your creature against the background you will quickly appreciate how important stillness is to the camouflage.

Sometimes the reverse is necessary and clothes must be seen.

When is it necessary to stand out clearly?

Which kind of clothes show up best?

Mountain Rescue

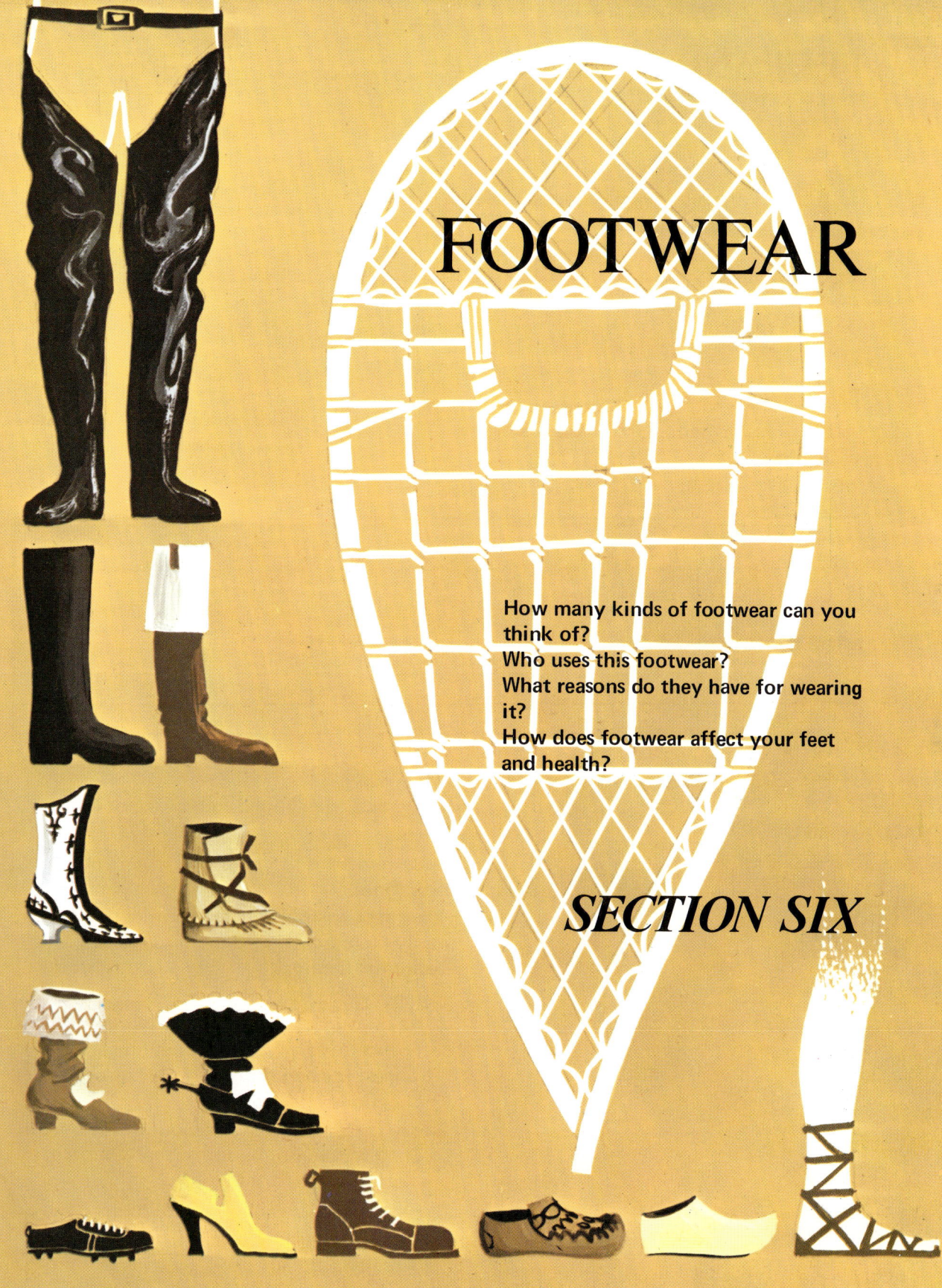

FOOTWEAR

How many kinds of footwear can you think of?
Who uses this footwear?
What reasons do they have for wearing it?
How does footwear affect your feet and health?

SECTION SIX

Top: Thigh boots. *Second Row:* Wellington boot, Hunting boot. *Third Row:* Fashion boot 1888, Alaskan Leather boot.
Fourth Row: Cavalier's boot, Boot 1638. *Bottom Row:* Football Boot, Fashion shoe, Boot, Moccasin, Clog, Roman Sandal.

FEET AND FOOTWEAR

How are shoes measured?

When you buy a size 4 shoe, what does this mean — four what?

To find this answer you will have to research into the story of measurement.

Make a survey of shoe sizes in your class or school.

Does the shoe shop want to stock equal numbers of all sizes?

From a survey could you suggest how many of each size would be required for a stock of a thousand?

Measure some feet:

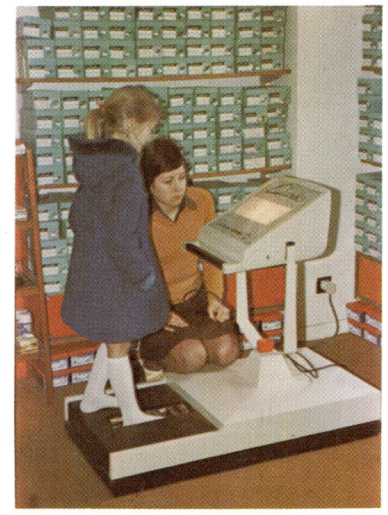

Electronic Foot Measurer

Personal information				Foot measurements								
Name	**Age**	**Mass**	**Height**	**Length**		**Width**		**Girth**		**Area**		**Shoe size**
				L	R	L	R	L	R	L	R	

a What patterns can you find?

b Is there any relationship between Age and Shoe Size, Height and Foot Area or Mass and Girth?

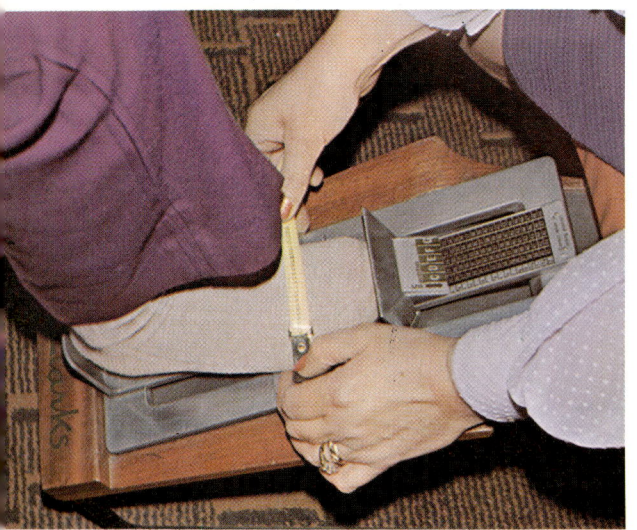

Foot measuring

Measurement is so important when you buy new shoes.

Your feet are growing and forming. They can easily be moulded into the wrong shape by badly fitting shoes.

Remember:-

1 Have your feet measured for your new shoes. They should be at least 1 cm longer than the longest toe and have plenty of room in all directions.

2 Try the shoes on, walk about in the shop. They should not be so big that your feet slip from side to side.

3 Party shoes are not made for long country walks.

Buy shoes for the job.

SNOWSHOES AND HIGH HEELS

Why do people in snowy climates wear snowshoes?

The owner of a rather fine floor once calculated that an elephant would do less damage to his floor than the lady!

It does seem that it makes a difference whether we spread force over a large area or concentrate it on a small area.

What downward force in NEWTONS do you exert?

Work out the force you exert on each square centimetre.

(This is called PRESSURE, and in science is more often measured in NEWTONS PER METRE^2)

Investigate this relationship between force and the area on which it acts.

Here is a method one school devised and used.

Why did this lady's high heels mark the floor?

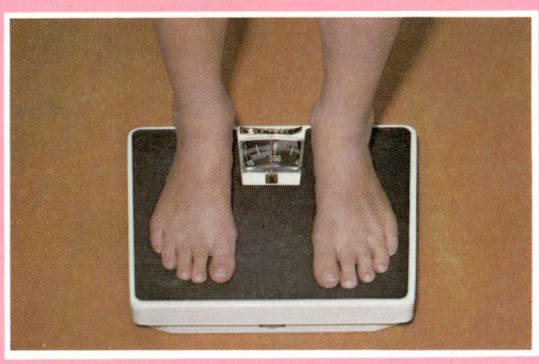

standing on Newton scales.

This is how they used it:
They found they still had a problem to measure the depth of the penetration.
How can you do this?

other shapes and sections to investigate

groove for string

5 cm

16 cm

plasticine

4 cm

8 cm

12 cm

support with chairs, etc

30 cm

block board or plywood

strings

bucket of sand represents the person's force acting downwards

using the test rig

SOLES AND SLIPPING

One thing we expect from our shoes is that we do not slip and fall over.

What sort of grip do different types of sole give on different surfaces?

Collect some soles
a rubber
b plastic
c leather
d worn examples

These will need to be pinned to a piece of wood.
(For a fair experiment think about the size of the soles and the load).

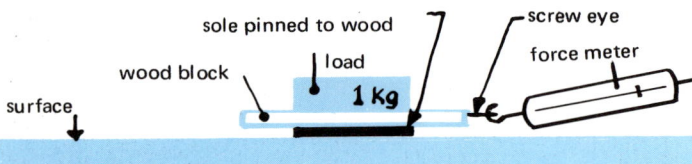

surface

wood block

sole pinned to wood

load

1 Kg

screw eye

force meter

How could grip be measured?
You will need a pulling arrangement that will measure the force required to start slipping.

Use some different surfaces

a linoleum   c cork   e plastic   g ice
b carpet   d wood   f concrete   h grass

Record your observations:

Sole	Surface	Force
leather	carpet	?

How does a climber get a grip on icy rock?

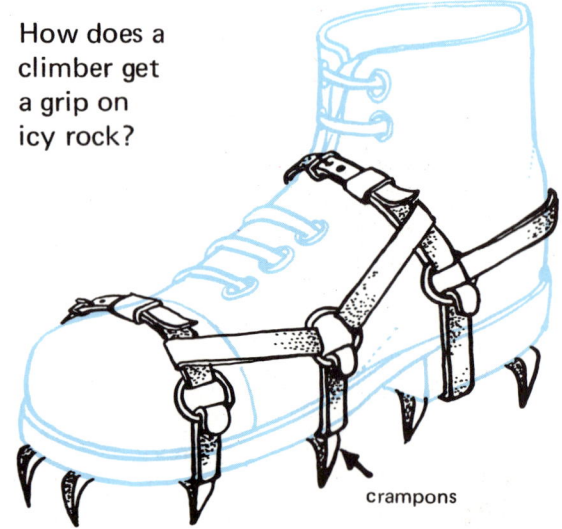

crampons

Do some further research into what happens if water is spilled on these surfaces.

From your results what have you found which will help you to choose the right footwear for the occasion?

40

Clothing & Electricity

Have you noticed when you are taking your clothes off that some of them have become charged?
Have you heard crackles?

Have they pulled towards other things or to you?
This could be a nuisance or even dangerous.

Here is one place where charges need to be avoided.
Can you think of others?

SECTION SEVEN

RUBBING AND CHARGING

Try rubbing different substances with various materials. See if they will then attract any small particles.

Here are some suggestions

a Things to rub:

Plastic rods; Strips of polystyrene ceiling tiles; Glass rods; Combs; Sticks of sealing wax; Metal Rods; Rubber; Pens; Cardboard; Wooden rods; Balloons; Strips of plastics.

b Materials to rub with:
Felt; Silk; Fur; Nylon; Wool; Cotton; Rayon.

c Small particles to test for attraction:

Polystyrene pieces; Cork chips; Pith balls; Tissue paper;
Metal foil; Fibres; Dust; Kitchen powders.

Which are charged and will attract?

What will they attract?
Does it matter what they are rubbed with?

To find a pattern of results you will have to record as you work.

Things rubbed	Material rubbed with	observation
Plastic	Felt	?

➡ ATTRACTING ⬅

Try some of the things you now know can be charged on other materials.

Here are some experiments for you to try.

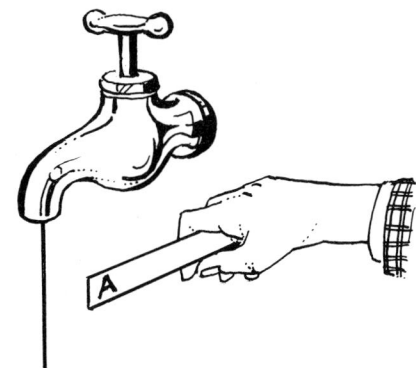

a hold charged material 'A' close to running tap

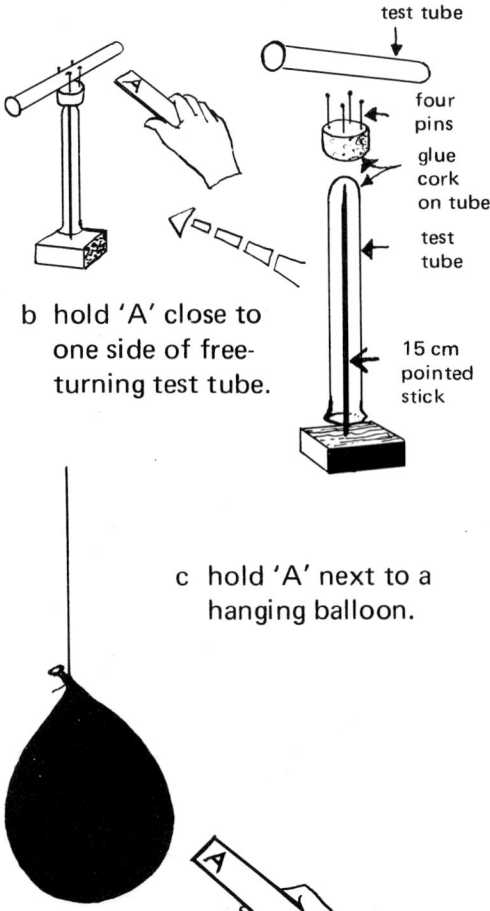

test tube

four pins

glue cork on tube

test tube

15 cm pointed stick

b hold 'A' close to one side of free-turning test tube.

c hold 'A' next to a hanging balloon.

⬅ AND REPELLING ➡

Is it always attraction?

Experiment further to try to answer this.

Charge two balloons by rubbing on a woolly jumper.

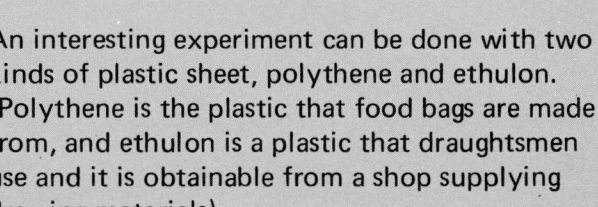

Bring them together

Charge a plastic rod by rubbing and suspend it in a cradle.

Bring a similarly charged rod near to it.

An interesting experiment can be done with two kinds of plastic sheet, polythene and ethulon. (Polythene is the plastic that food bags are made from, and ethulon is a plastic that draughtsmen use and it is obtainable from a shop supplying drawing materials).

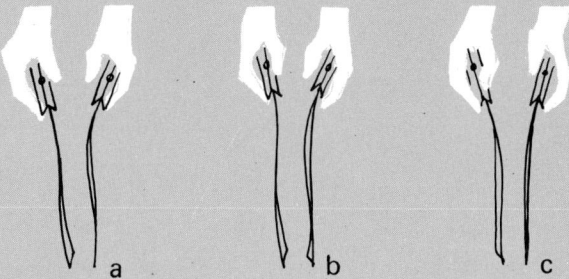

a   b   c

Charge the strips by rubbing them between your fingers.
Bring together:
a two strips of charged polythene
b two strips of charged ethulon
c charged polythene and charged ethulon

Can you find any rule about charged materials that are attracted and charged materials that are repelled?

STATIC ELECTRICITY

This effect you have experimented with by rubbing is known as STATIC ELECTRICITY.

It was the first kind of electricity that was noticed. It was discovered by the ancient Greeks who found that amber when rubbed by fur would attract small particles. The Greek work for amber was ήλεκτρον (elektron).

Scientists use an instrument called an electroscope to test whether an object is charged.

above: electroscope

left: a young scientist experimenting with an electroscope

You could try making your own electroscope from easily obtainable odds and ends.

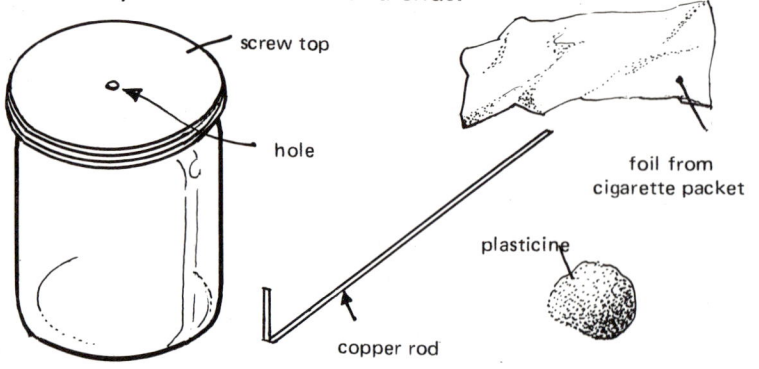

screw top

hole

copper rod

foil from cigarette packet

plasticine

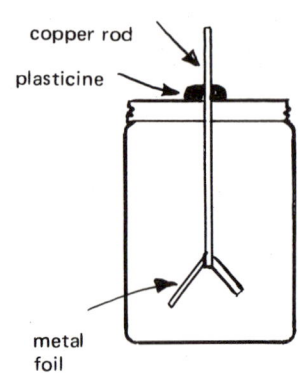

copper rod

plasticine

metal foil

CLOTHES AND ELECTRICITY New discoveries claim to reduce the 'cling and crackle' of man-made fibres.
Courtaulds 'Anti-Stat Celon' is a new cling-resistant, crackle-free, dirt-resistant fabric.
Also, we can now add a special rinse to the wash to reduce these unwanted charges.
Can you test any of these claims?

Cleaning and caring for clothes

Our clothes become soiled. They must be cleaned.
a How are clothes washed?
b Is hand washing better than machine washing?
c How do washing machines work?
d When is soaking necessary?
e Which washes best — soap, washing powder,
   biological powders or liquids?
f Are different kinds of materials washed differently?

STARCH

BLEACH

WHITER AND BRIGHTER

Look at the claims made for the various washing powders.

What claims are made in television advertisements, in the press and on posters? What do mothers say?

Conduct a scientific enquiry into washing powders.

Washes whiter

Removes all stains

Adds brightness to whiteness

Here is how two young girl scientists carried out such a test. They were concerned that the test should be fair.

Their test was to be on equal-sizes pieces of white cloth, stained with equal amounts of ink, fruit juice, oil and tea. They decided they must use equal amounts of water and that it must be the same temperature for each wash.

Equal amounts of each washing powder must be used.

They also agitated (stirred as the washing machine does) the wash for equally timed periods.

Do you agree that the test was a fair one?

Even when they had completed their washing test, they were still faced with a problem.

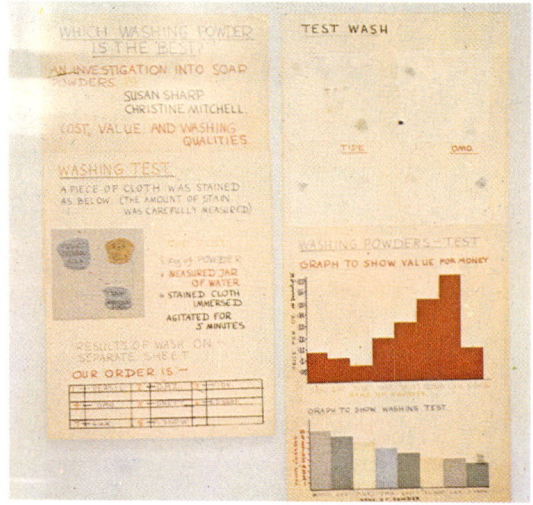

Washing Powder Test Results

Which was the whitest?

How do you think they solved this problem and presented the wash in an order of whiteness?

While they were working two friends noticed that one powder claimed to give more value for money.

How could this be investigated?

Collect the various powders and try this investigation for yourself.

SOAPS AND DETERGENTS

You could carry out some further investigations into washing and using soaps and detergents.

Stain and dirty some squares of material and wash them in different ways.

a Wash in cold water only

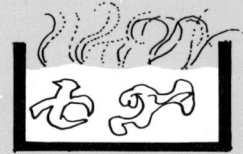

b Wash in hot water only

c Wash in cold water plus soap powder

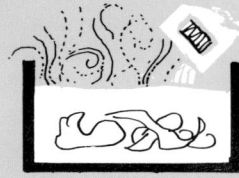

d Wash in hot water plus soap powder

In each case do two types of washing:

i Soak and lightly squeeze.
ii Agitate violently for the whole wash.

Which combination seems best?
Is it the one used by washing machines?

Why do mothers and laundries use soaps when washing clothes? "To get the clothes clean", is the obvious answer, but how do they help to remove the dirt? What does the soap do?

Try these experiments with soap. They will help you to find answers for yourself.

1

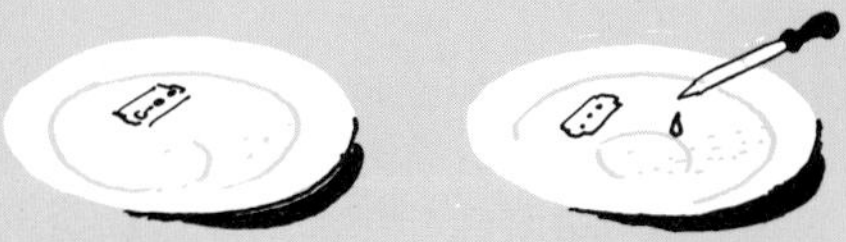

a Float a razor blade in a saucer of water

b Add a drop of liquid soap or detergent

2

plain water

soapy water

sprinkle a little black powder (soot; powder paint or dust) onto the surface of each — observe.

3

plain water

soapy water

Float tiny squares of cotton (1 cm^2) cloth, in each jar. Time the cloth getting wet and sinking.

4

glass rod

dropper

yogurt pot

a Place a piece of gaberdine over a pot and drop water on it.

b Add a few spots of detergent to the water drops.

5 Try placing drops of water onto various surfaces.

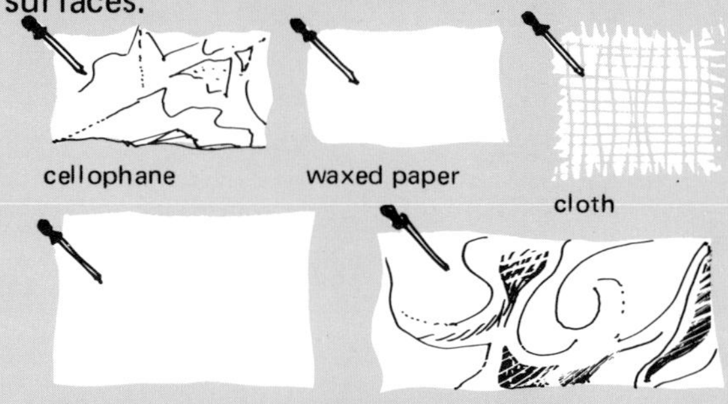

cellophane

waxed paper

cloth

plastic

aluminum foil

Repeat this with water which has a little detergent added.

You could also try different kinds and strengths of soap or detergent and hot and cold water to see if any differences can be observed.

WET WATER

You have been observing differences when soap or detergent is added to water. You will find it a challenge to try to measure this difference.

Water particles have forces acting between them which hold them together. The effect of these forces at the surface can be seen and measured.

Such forces pull drops into rounded shapes and support things like Pond Skaters.
This is called surface tension.

A simple balance could be used to measure surface tension.

Here is one made by a class for this purpose.

Pond Skater

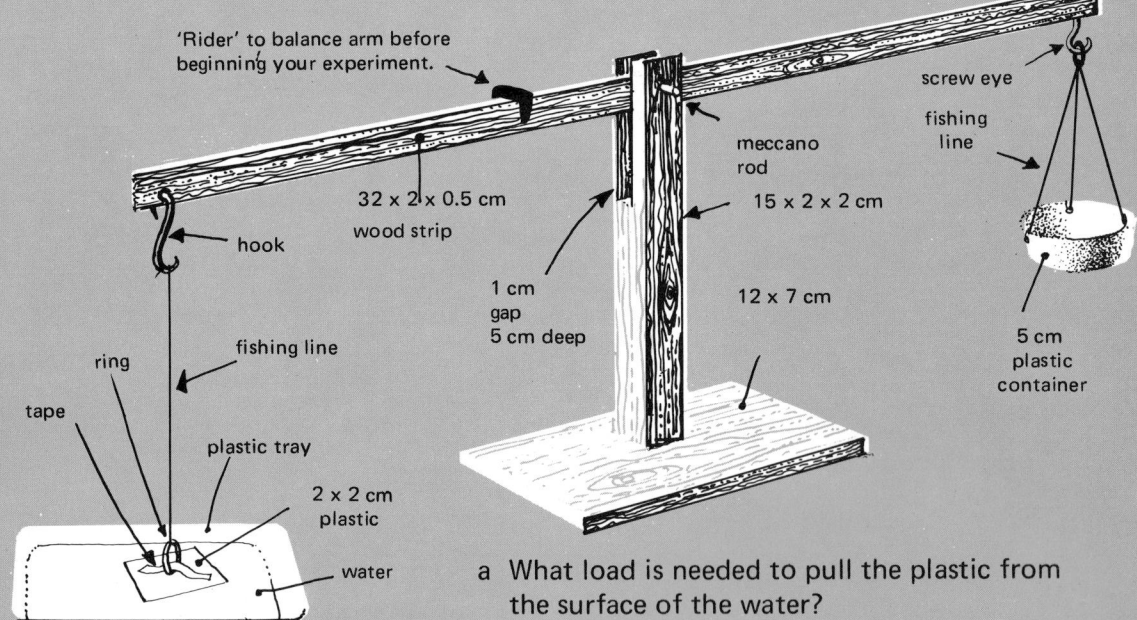

'Rider' to balance arm before beginning your experiment.

screw eye

fishing line

meccano rod

15 x 2 x 2 cm

32 x 2 x 0.5 cm wood strip

hook

1 cm gap 5 cm deep

12 x 7 cm

5 cm plastic container

ring

fishing line

tape

plastic tray

2 x 2 cm plastic

water

c Repeat your experiments but now add drops of detergent to the water.

a What load is needed to pull the plastic from the surface of the water?

b Experiment with different shapes, perimeters and areas. *(dimensions are in centimetres)*

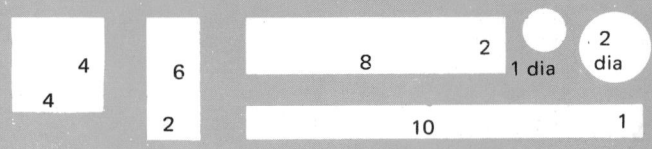

Soap and detergent reduce the surface tension and so increase the water's power to spread and run. Soap makes water 'wetter'.

HARD AND SOFT WATER

In some places it is much more difficult to obtain a lather than in other areas. The people who live in such places say that their water is *hard* — hard to obtain a lather with soap. Those whose water is easy to lather call it the opposite, *soft* water.

With the help of teachers and friends you should be able to obtain samples of water from widely differing areas of the country.

See also that you have the two extremes.

a Distilled water — (soft water).

b Tap water plus magnesium sulphate (Epsom salts.)   — (Hard water artificially made).

How can the hardness of water be measured? Add 'one' soapflake at a time, to a little water in a test tube.   Shake the test tube — until a head of lather remains for one minute.

It would be good to record your findings on a geological map of the country, as you might then be able to see a pattern that related your hardness and softness to other things.

Record the number of soap-flakes used as a measure of the hardness.

WASH DAY INVESTIGATIONS

What other chemicals are used when washing clothes?

Soda, starch and bleach are three common ones.

Why are they used?

Soda (Sodium Carbonate Crystals).

This is used to 'soften' the water. You could easily test this by repeating the experiments on page 49. Use the hard water, but this time adding some soda and comparing results.

Starch

This is used to stiffen a fabric and give it some firmness.

Here are some different kinds of starch.
a Cold water starch
b Maize starch
c Instant Starch
d Plastic Starch
e Spray-on Starch

Experience mixing and using starch for yourself.

Mix some starch of different strengths.

Treat some samples with each solution.
Which strength would you recommend for:-
i   An Elizabethan ruff?
ii   A collar?
iii A blouse?

Bleach

This is used to whiten fabrics (it is advisable to wear rubber or plastic gloves and an overall when working with bleach. Be careful not to splash but if any does get on your skin wash immediately under a running tap).

What colouring will bleach remove?
a Spot some on to various fabrics
b Add some drops of bleach to these coloured solutions.

WASHING MACHINES

Find out about washing machines.

What types are there?

How do they work?

Collect some advertising leaflets and find out about various types and compare methods, claims and prices.

What different ways of moving the water and clothes are there?

How long have we had electrical washing machines? What was used before?

What are these, how were they used?

Rocks and Water

'Dolly'

Mangle

How is washing done in some of the less developed areas of the world?

Can you invent a model washing machine?

Many washing machines have a spin dryer. How does this remove water from the washing?

You could make a model spin dryer from a squeezy bottle and a piece of string.

19th century copper

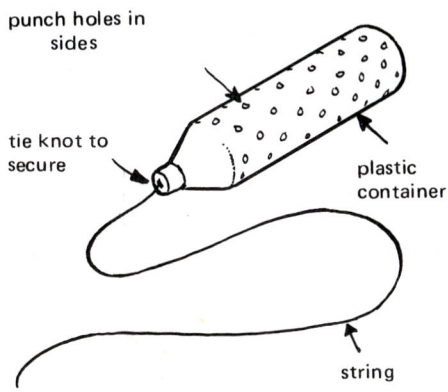
punch holes in sides

tie knot to secure

plastic container

string

Carry out some tests with a washed handkerchief.

What difference to dryness does the number of times whirled make?

What difference does the number of holes, or the size of holes make?

Get your friends to make some different 'spin dryers' and compare.

How will you be fair and make sure that each handkerchief is equally wet? How will you measure how much water has been whirled away?

IRONING

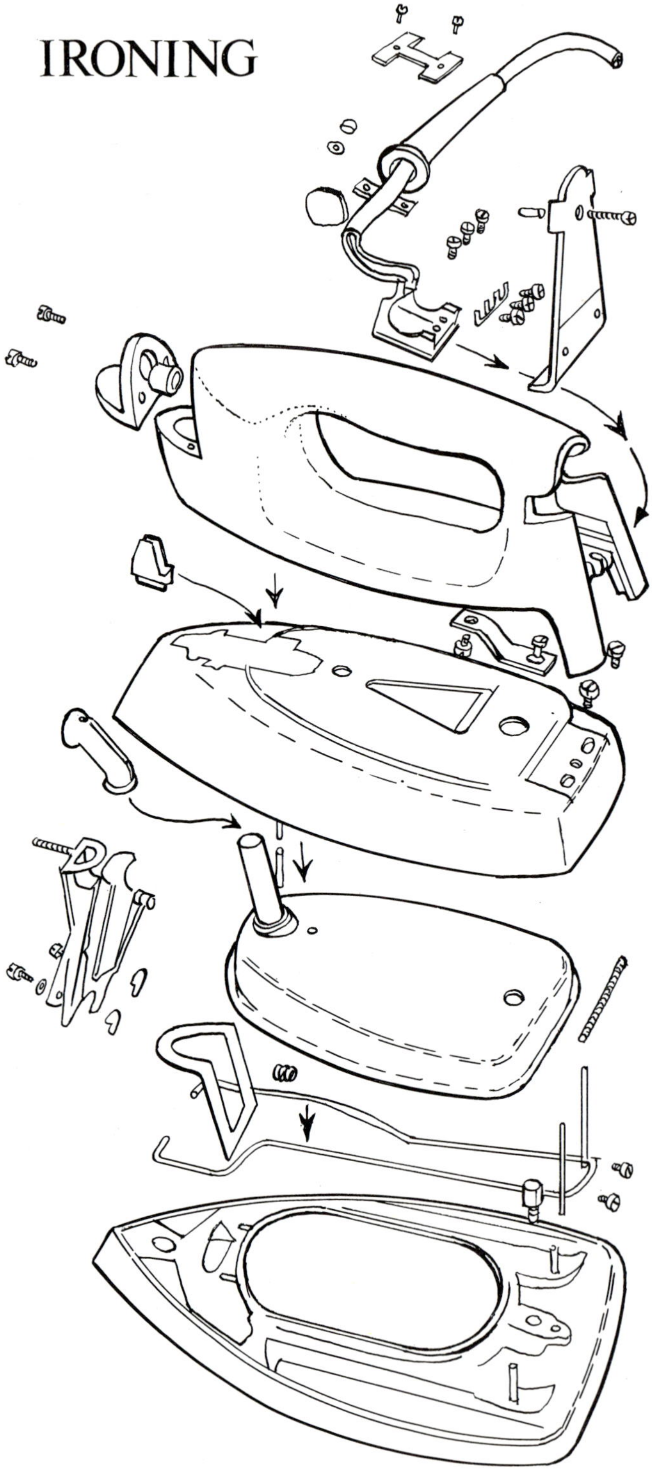

Early steam and flat irons

To keep our clothes smart and looking well cared for, they are pressed.

Shirts are ironed.
Trousers are creased.
Pleats are pressed.

This is done by using a hot iron.

How is the iron heated?

Not many years ago they were heated on a fire or gas ring.

Investigate electricity and heat.
(If you have not explored electricity before you will need to find out about circuits and switches, conductors and insulators, which you can do from Section of *Houses and Homes.*)

A model electric element.

Start your investigation with 10 cm, 34 swg nickel chrome wire.
Investigate by varying the length, thickness and kind of wire.

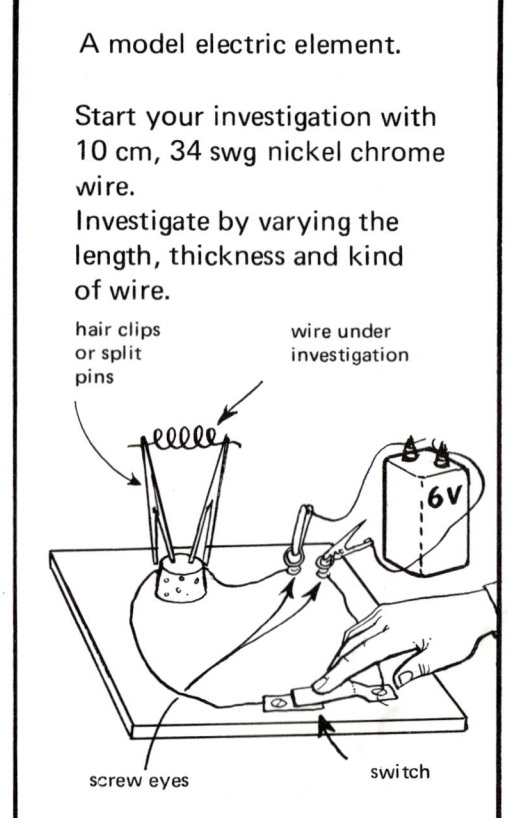

hair clips or split pins

wire under investigation

6V

screw eyes   switch

Today we use the electric iron. Try to obtain an old iron and take it to pieces.
Here is an exploded view that will help explain what you find.

Dyes and Dyeing

During your research into clothes of past ages have you thought how their colours were obtained?

Can you find from what these dyes were obtained?

Murex shell

A Roman emperor. His toga was dyed with the colour from the murex shell.

An ancient Briton. His tunic is dyed with woad, producing a blue colour.

Woad

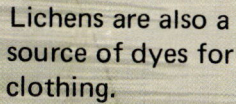

Lichens are also a source of dyes for clothing.

Saffron Crocus

A Buddhist monk.
His cowl is coloured with saffron, which comes from the orange coloured stigma of the mauve flowering crocus, native to Asia and part of Europe. It requires half a million stigmas to make one kilogram of colour.

A Grenadier Guard - 1845
The uniform was dyed red with cochineal, the dried shells of the fesal cactus insect of Mexico.

Cochineal insect (5 times actual size)

SECTION NINE

53

OBTAINING DYES

On a woodland visit or a discovery walk have you wondered if the colour could be extracted from berries, flowers, barks, lichens, roots and brackens?

Can the colour from natural materials be used to dye cloth?

Some suggestions for you to try are:-

Red cabbage, Pea pods, Tomato leaves, Beetroot, Onion skins, Lichen, Tea, Coffee, Dandelion roots, Elderberries, Blackberries and Lily-of-the-Valley leaves.

(Remember - some berries and leaves are poisonous.)

How to start

These children chopped up the material and left it to soak overnight. They pounded and squashed the mixture, then heated and let it boil for about fifteen minutes.
They strained off the liquid when it was cool.
They dipped their piece of cloth in this liquid, the dye, and let it simmer for a few minutes.

left: Children preparing dyes

If possible obtain some natural samples of wool, cotton and linen and also pieces of rayon and nylon. See how each fabric takes the dye.

Does the time in the dye make any difference.

Does the temperature make any difference?

Try while the dye simmers on heat.

Try leaving the dye to become cold before using.

Does the water make any difference?

Try using rain water, tap water, soft water, hard water.

(Find out why the streams of the Pennines were important to the cotton and wool industries of Lancashire and Yorkshire.)

FIXING AND FASTNESS

a Does your dye wash out?
b Does your dye fade in the sunshine?
c In other words, is your dye *fast?*
d Are some dyes fast only on some
    fabrics?

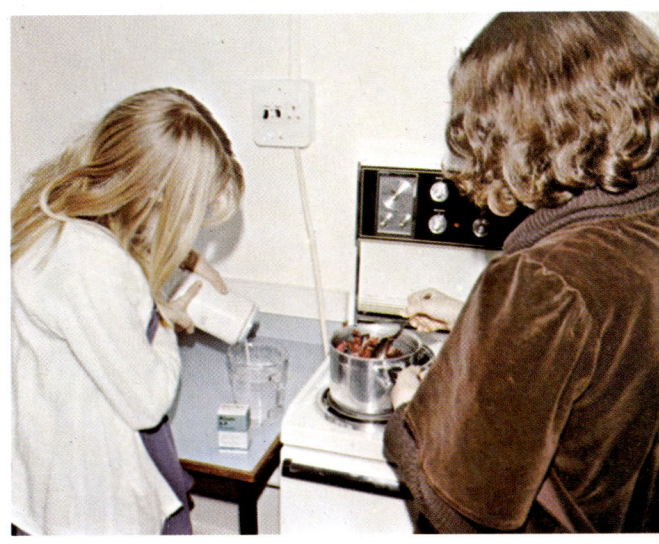

Two girls experimenting with dyes, this time adding a mordant.

Certain chemicals have been and are used
to help the dye to fix on to a fabric and
so make the colour fast. These chemicals
are called *mordants.* (The word mordant
comes from a Latin word 'mordere' which
means 'to bite'.)

You could investigate with some mordants
and see what effect they have.

Some to try are:—
a Common Salt (Sodium Chloride).
b Alum (Potassium aluminium sulphate).
c Green Iron Sulphate (Iron II sulphate).

Mix your mordant with water to make a solution. Soak the sample for a few moments in
this solution and then dye as before.

Fascinating colour changes can also be found when investigating mordants.

Here is how they recorded their work.

How will you test to see if the dye is now
fast?

Do different fabrics behave differently with
different mordants?

Try using the same fabric and the same dye
but use different mordants.

Try using the same dye and same mordant
but different fabrics.

This chart shows one group's results.

DYEING

You will probably like to compare some of your natural dyes with the commercial synthetic (artificial) dyes.

Find out about Sir William Henry Perkin and how he discovered the first synthetic dye, aniline purple.

Repeat some of your first experiments again. Use the various types of fabric and see if you are more successful with synthetic dyes on those fabrics that did not seem to take a natural dye.

If you are interested in art and craft work you will like to try some of the dyeing crafts:-

a Tie and Dye
b Batik
c Bleach remove
d Fabric printing

It is interesting to look at the names given to the colours by manufacturers. Can you make up names for your dye colours?

Above: Fabric printing. *Top right:* Bleach remove
Bottom right: Tie and Dye.

CLOTH CLAIMS

COATES COTTONS
NON-FLAMMABLE

HARD WEARING
LOCHII
HARRIS TV

wool
SHRINK
RESISTANT

Manufacturers advertise
such claims as these
What do they mean?
CAN THEY BE TESTED?

THORN PROOF
Glengarrif
IRISH TWEED

CREASE
RESISTANT

NON IRON

drip dry
NYLO

MOTH
PROOF

SECTION TEN

WEAR AND TEAR

We expect some of our clothing to stand up to hard wear:- school uniforms, overalls, playclothes and work clothes for instance.

Investigate hard wearing fabrics
School clothing, trousers and blazers, jeans, overalls and tailors' samples will provide small pieces for experiments. (Worn out and discarded items will have parts that are unworn that can be used.)

Here is a simple test:-

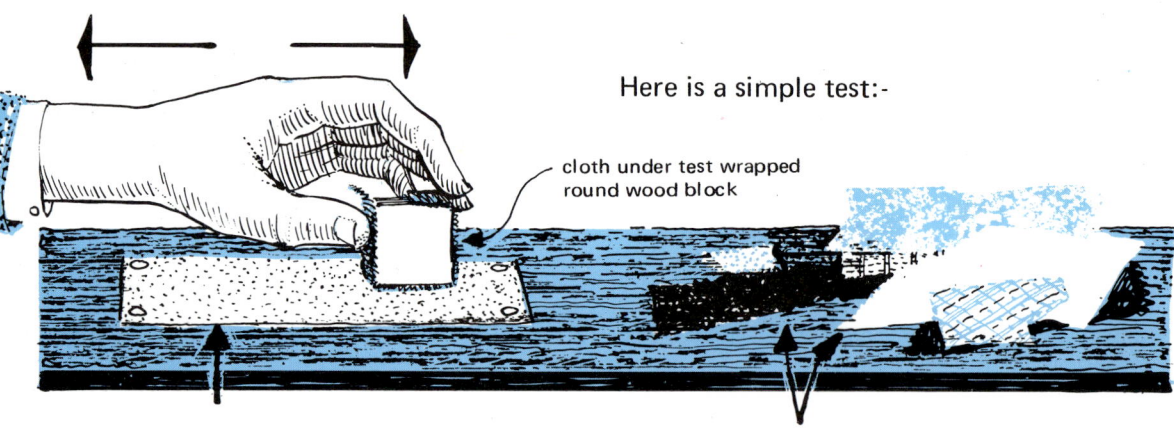

cloth under test wrapped round wood block

glasspaper secured to board

cloth pieces for test

Rub once, five times, ten times and examine with your lens. Record your observations.

Have you noticed leather being used on the elbows and cuffs of jackets to save wear and tear? Repeat your experiment with a piece of leather. How does this compare with your fabrics?

THORN PROOF   Obtain some samples of 'thorn proof' materials, plus a selection of other fabrics.
Carry out a tear test.

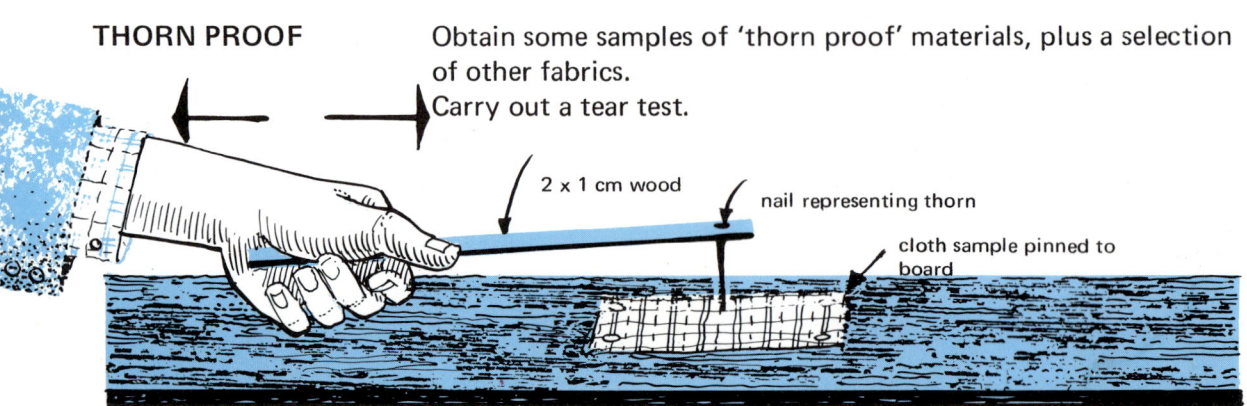

2 x 1 cm wood

nail representing thorn

cloth sample pinned to board

Which types resist a thorn tear the best? How did the 'thorn proof' sample fare?

MOTH PROOF

Moth Proof' is an interesting claim. Why do clothes need to be moth proof? What do moths do to clothes?

Can you collect some information about this? Perhaps you can find out if the moth attacks any particular kind of clothing.

Here are some drawings of the life story of the clothes moth. (Tineola bisselliella.)

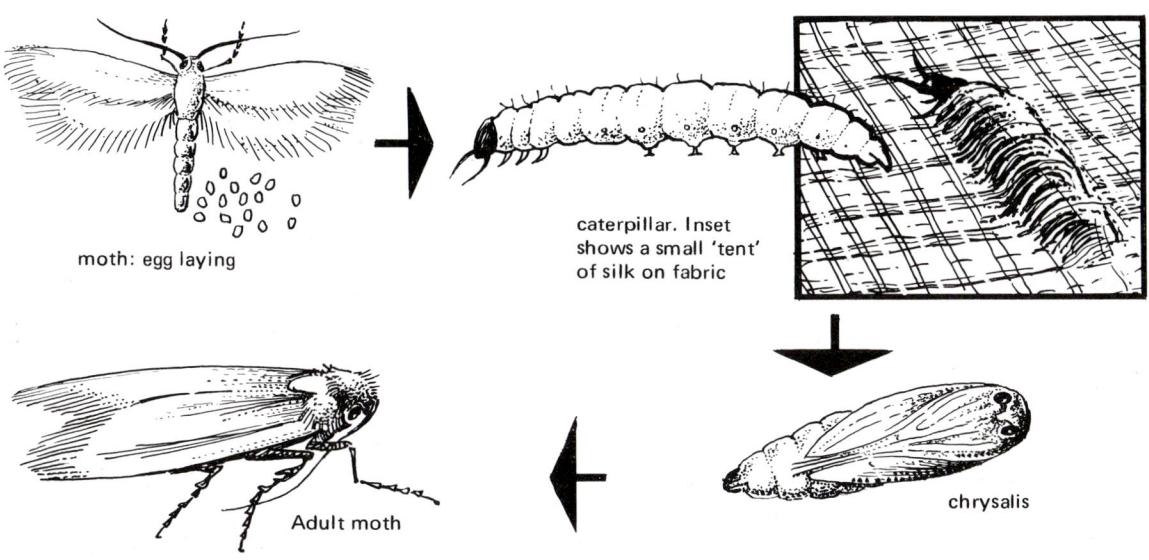

moth: egg laying

caterpillar. Inset shows a small 'tent' of silk on fabric

chrysalis

Adult moth

You might think that clothes would not make a very good meal. There are, however, a number of insects that feed on such materials. We think they first fed on bits from birds' nests and small animals' homes. They have now moved to our homes and are rarely found away from this rich supply of food.

It is the caterpillars that do the damage and they feed on woollen things, furs and feathers. They eat night and day and so soon make a hole. Like other caterpillars, they change their skin three or four times before changing into a chrysalis.

These insects like the dark and being left alone. When disturbed the male moth flies but the female tries to run away. Her body is full of eggs and so too heavy to fly easily. Clothes which are often used are therefore not likely to be a clothes moth's meal.

Other ways of keeping the moths away include storing the clothes in plastic bags and the use of chemicals. Can you find which chemicals are used? Unfortunately most of these do not have a lasting effect. A modern idea is to make the cloth so that the caterpillar does not like to eat it. This is the best way and such materials are called moth proof.

FLAME Resistant ?!

There is a very real danger of injury by burns. These can result from fabrics that flare and catch fire easily. The danger is greatest with night garments and open fires.

(Do be careful; remember that the fire's draught draws loose clothing in towards the fire. So help safeguard younger brothers and sisters from such dangers).

Design some posters showing where such danger exists: open fires, electric fires, oil stoves, party dresses, nightwear and fancy dress.

Carry out some research into fabrics and how easily they catch fire.

This is the time for you to talk with your teacher about the dangers of burning and the safety precautions that should be taken.

Apparatus for burning test. Tray with sand, 'meths' burner, tweezers.

Testing a sample cloth.

Obtain some material that is said to be flame resistant. Test it with some material that is not flame resistant and compare.

Experiment in trying to make a fabric flame resistant.

Try soaking your test piece in various solutions.

Some you might try are:-
a Salt (Sodium Chloride)
b Alum (Potassium Aluminium Sulphate).
c Borax (Sodium Borate).

When dry, burn and compare with non-treated samples, taking the same safety precautions.

Clothes can be more than a covering

What are these people wearing that is more than just a covering?

Why do you think they choose to use these things and dress like this?

If you look at costume through the ages, or just at the fashions in your local High Street, you will see there is more to clothing than keeping warm and dry.

SECTION ELEVEN

DECORATION AND ADORNMENT

What is used to make clothing more attactive?

Some adornments are:- buttons, ribbons, jewellery, brooches, cuff links, flowers real and artificial, buttonholes and scarves. What others can you think of?

What are the raw materials used for these decorative additions to our dress? Where are they obtained?

Decorative feature	Source of raw material	Country of origin
Cameo brooch	Sea shell	Italy
Gold necklace		
Plastic button		
Tiger's eye cuff links		
Diamond ear-ring		
Nylon ribbon		
Silk scarf		
Silver tie pin		
Pearl necklace		
Coral bracelet		

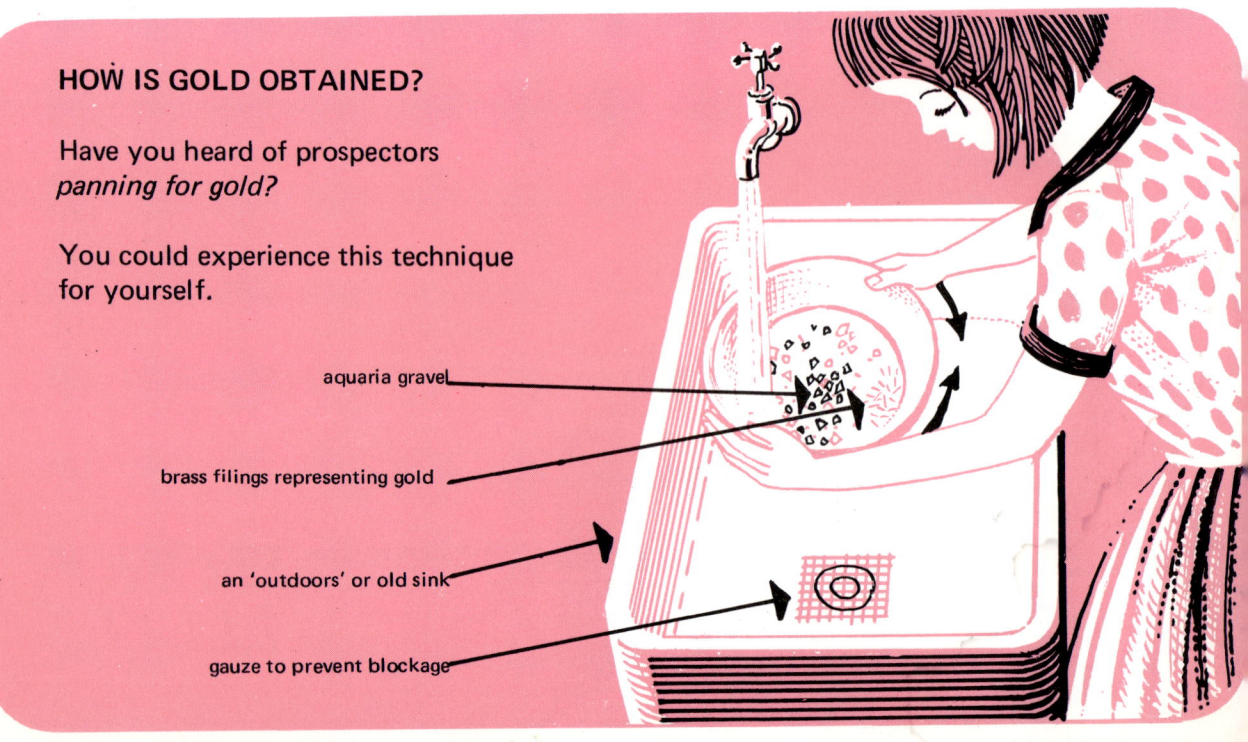

HOW IS GOLD OBTAINED?

Have you heard of prospectors *panning for gold?*

You could experience this technique for yourself.

aquaria gravel

brass filings representing gold

an 'outdoors' or old sink

gauze to prevent blockage